Plant Cell Biology

Plant Cell Biology

Virendra Batra

Oxford Book Company
Jaipur ▪ New Delhi

Published By

Oxford Book Company

Regd. Off. "Bony Residency", Gate No. 2
Opp. Tilak Public School, Vishweswariya Nagar,
Gopalpura Road, Jaipur - 302018, India
Ph: 0141 - 2761280, Telefax: 2761381
email - oxfordbookjpr@gmail.com
website - oxfordbookcompany.com

Delhi Office: 102, 1st Floor, Satyam House,
4327/3, Ansari Road, Darya Ganj,
New Delhi - 110002, India
Ph: 011-45652440

Distributed By
KEYA INTERNATIONAL PRESS LLP
E-mail: keyainternationalpressllp@gmail.com

Published in India, 2024

ISBN : 9788189473648

Price: ₹ 1950.00

Typeset by Shivangi Computers, Jaipur

Printed and bound in India by Thomson Press Ltd.,
New Delhi

Preface

The rapid advances in sciences such as biotechnology and genetic engineering, atleast within the context of plants, has been given immense incentive with the progress in cell biology, which undertakes a comprehensive study of cell structure, functioning and pathology. With the development of such instruments as the electron microscope and the scanning electron microscope, the scope of plant cell biology has been greatly widened to understand and incorporate knowledge within fields like biochemistry and cytogenetics. As such the relevance of this field cannot be overemphasised, especially as it offers immense prospects in medicine, agriculture and so on.

The present text has been aimed as an introductory manual on the subject of plant cell biology, seeking to acquaint botany students with the fundamental principles and applications of the field. Delineating the concepts, techniques and processes associated with plant cell biology, the text attempts to highlight the role of plant cell biology in shaping biochemistry and biotechnology. Emphasising upon the ways in which cell biology has expanded the range of botanical studies, the text analyses current trends and developments, and offers insight into the future of the field.

Virendra Batra

Contents

1

Introduction to Microscopy

For many biologists, the microscope belongs to the gadgets they use most. The reason is obvious: cells are the basic units of life and their size is, with only few exceptions, below the upper limit of the resolution of the human eye. Numerous statements on the refraction of light by lenses or waterfilled spheres and the resulting magnification proof, that the effects of magnification were already known in antiquity. In the 13th century, the British natural scientist Roger Bacon used segments of glass spheres as magnifying glasses and recommended them to partially sighted persons as spectacles. The grinding of lenses was perfected in the following centuries.

The physician P. Borch records, that Hans and Zachaias Jensen (father and son) from the Dutch town Middelburg invented the compound microscope at the beginning of the 17th century by placing one lens behind the other. They are regarded as the inventors of the telescope, too. Independent of the beginning development of the two-lensed—also called the compound microscope—stronger and stronger lenses became fashionable, too. Again, it was a Dutchman, A. vanLeewenhoek, who succeeded in the construction of lenses with 270 fold magnification and a focal length of about one millimetre, with which he worked. His resulting observations constitute the pinnacle of "single-lensed microscopes".

In 1667, British natural scientist Robert Hooke examined vegetable tissues with the help of a compound microscope and thus detected their basic elements, the cells. The further development of microscopes and especially the use of microscopes in the 18th century went on rather hesitantly. This was on one hand caused by the opinion of many natural scientists of reputation, that a lot had still to be discovered even without the use of microscopes. On the other hand, the interfering chromatic aberrations were a continuing source of misinterpretations. This led .to the prejudice, that anything you wanted to see could be seen within a microscope.

As soon as 1695, D. Gregory considered the combination of several lenses with differing dispersions in order to minimise this handicap. The mathematician L. EULER brought forth a theoretical explanation of achromates (systems of lenses with corrected colors) in 1771 and he suggested the calculation of such achromatic objectives. They were finally built by F. G. Beeldnyder (1735 - 1808), who placed a biconcave lens out of flint glass between two biconvex lenses out of another type of glass, that had a different index of refraction.

G. D. Amici developed achromatic systems of lenses further and invented also immersion objectives. The actual break-through on the way to the construction of modern microscopes was due to the formulation of a theory of image formation within the microscope by E. Abbe. He proved, that an absolute upper limit of microscopic image formation exists and that it was dependent of the aperture of the objective and the wave length of the light. His theory permitted the development of reproducible high-performance microscopes and their production in series.

The progress of microscope building enabled a quick series of numerous discoveries in the subjects of histology, cytology and bacteriology. The progress was helped by the development and use of suitable methods of fixation, embedding and cutting, specific dyes and conservatives.

THEORY OF LIGHT MICROSCOPY

The closer an object is to the eye, the more details can be distinguished until the limit is reached, below which the eye cannot depict the object clearly any more. This limit is 250 mm for the average grown-up. The measure for the ability to tell two points apart is called the power of resolution. It describes whether two adjoining points can still be perceived as separate. To examine objects below this limit of resolution, a magnifying glass is needed. Its magnification can be calculated in the following way:

$$V = 250\ [\text{mm}]\ /\ f\ [\text{mm}].$$

250 mm is the image distance of the human eye and f is the focal length of the lens. If the magnifying power of the lens is known and the focal length has to be calculated, than the formula has to be used as shown beneath:

$$f\ [\text{mm}] = 250\ [\text{mm}]\ /\ V$$

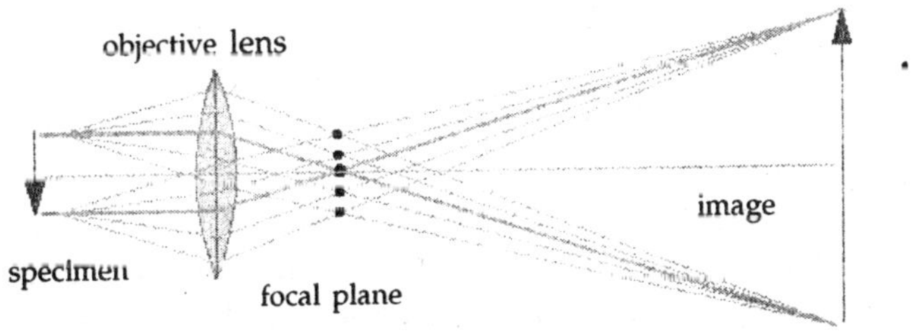

The picture shows the way the fan of rays of an object that is collected by a single lens is imaged.

If you want to magnify an object even more, two lenses in tandem have to be used (objective and eyepiece or ocular). The resulting construct is a simple microscope. The objective magnifies the object or specimen (O) and a turned up real image (O′) is formed in front of the focal plane of the second lens. The eyepiece than forms an enlarged virtual image (O″), that can be seen as a turned up image at a distance of 250 mm. The magnification of a microscope is thus a product of

$$V_{objective} \times V_{ocular}$$

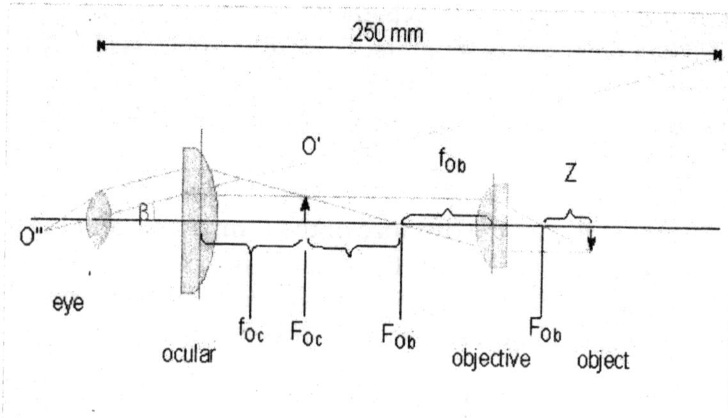

Figure 1. Light path in the microscope: F = Focal plane, O = Object (Specimen), Ob = Objective, Oc = ocular (eyepiece)

Nonetheless the power of resolution is given exclusively by the objective. To describe an objective not only the magnifying power has to be taken in account, but, too, the numerical aperture. The numerical aperture is defined as the sine of half the angle of the cone of light from each point of the object, that can be accepted by the objective *(alpha)* multiplied by the index of refraction of the medium in which the object is immersed (n):

$$A = n \sin alpha.$$

The medium is usually air with a refraction index of n = 1. *alpha* can never be bigger than 90° and thus the numerical aperture can never outgrow 1. Its largest actual size is 0.95, since the distance between objective and the surface of the cover glass cannot reach zero. The aperture of 0.95 corresponds to an angle *alpha* of roughly 72°. An increase of the numerical aperture can be achieved by the choice of a medium between objective and object with an index of refraction bigger than that of air. Special oil for immersion with an index of n = 1.515 has proved to be useful. Larger indexes of refraction do not make sense, because the index of refraction of the objective itself (n = 1.525) becomes limiting. Immersion oil can be used only with specially constructed immersion objectives. If *alpha* has the maximum of 67.5°, the aperture is accordingly 1.515 x 0.92 = 1.40. The degree of resolution (d) is set by the wavelength of light *(lambda)* and the numerical aperture (A_{obj}):

$$d = lambda \ / \ A_{obj}$$

If *lambda* is 550 nm (green light) the formula runs the following way:

$$d = 550 \text{ [nm]} \ / \ 2 \times 1.40 = 200 \text{ nm} = 0.2 \ \mu\text{m}$$

0.2 µm is the highest theoretical resolution that can be reached with a light microscope. A rough approximation shows that the power of resolution of a light microscope lies at about half the length of a lightwave if a good immersion objective is used. If the limit of resolution of a microscope is known, then the maximal useful magnification can be calculated. A magnification is called useful when two only just clear points are magnified so strongly that they are seen as separate unities by the human eye. At 250mm is the resolution of the human eye about 0.15 - 0.2 mm. The rule of thumb for a useful magnification is thus:

$$500 - 1000 \times A_{obj.}$$

An objective with an aperture of 1.4 has accordingly a maximal useful magnification of 1400fold.

WAVE-PARTICLE NATURE OF LIGHT; PHENOMENONS OF DIFFRACTION AND IMAGE FORMATION

We assume that the dual nature of light as both wave and particle is known and we will in the following deal with some fundamental principles of wave optics in order to be able to interpret microscopic image formation. Our main interest, nevertheless, is the understanding of modern techniques of light microscopy like phase-contrast microscopes, polarising microscopes and interference microscopes. First some terms and definitions:

Amplitude, frequency, wave length, interference, diffraction and phase are the most important parameters to describe a wave. The wavelength of visible light is between 400 and 800 nm. Interference is the mutual influence of two waves on each other, whereby the resulting crests may be either enhanced or flattened (enhancement of amplitude, reduction of amplitude). The extreme case is given by two waves that extinguish each other.

Diffraction is the partial deflection of a ray of light at the corners of opaque objects. For understanding it is useful to cast a look at phenomenons of diffraction at simple gap- or hole-shaped masks. If a screen is placed behind such a mask and light is sent through, the pattern of diffraction can be seen. It consists of a regular pattern of points or lines that fades away towards the margins. The shape of the pattern is dependent on the kind of mask and the object used (for example a microscopic preparation). But the pattern is always grouped around a central axis (unmagnified picture of the gap or hole of the mask). If a mask with two openings is used, the light is diffracted at the rim of both openings. The diffraction images of the two rays overlie each other and regular enhancements and reductions can be seen in the resulting pattern.

To get a magnified image, the first additional maxima of the diffraction image have to be included. The results in the pictures above have already shown that the distances of the points in a diffraction image are reversed proportional to the corresponding distances in the object. To gain a real image from a diffraction image a collecting lens (the objective of the microscope is one) is needed. This lens has to have a big enough aperture to collect the intensity maxima (-I, +I). The real image is formed by both refracted and non-diffracted light.

Magnifications as those that have been described can, too, be achieved with radiation of other natures. The most important additional device for biologists besides the light microscope is the electron microscope, where the diffraction of a ray of electrons is used. Another radiation to be mentioned is the short-waved Röntgen-radiation (X ray-radiation), that has proven to be of value for the elucidation of molecular structures. Since no collecting lenses for Röntgen-radiation exist, the information contained in the diffraction pattern has to be calculated mathematically in order to get to know the structure of the molecule.

PHASE-CONTRAST AND DARK FIELD MICROSCOPY

A main obstacle in the microscopy of biological objects is their poor contrast. Only where a contrast exists or where it can be achieved by contrast-enhancing dyes, structures can be made visible. Light-absorbing parts of a preparation weaken the amplitude of the light waves that pass through them. It is thus also spoken of amplitude preparations. The change of stronger and feebler light is perceived by the eye as a difference in brightness. The invisible parts of the preparation are went through by the light without a change of amplitude, but the phase of the light may be changed depending on the consistency of the material. This change is due to the altered speed of the lightwaves. Differences in phase can be perceived neither by the eye nor by a photographical film.

The Dutch physicist F. Zernike succeeded in 1935 to convert phase to amplitude differences. He was awarded the Nobel price for this achievement in 1953. Today his method is known as phase-contrast microscopy and it is by now an integral part of nearly all research and many teaching microscopes. Its eminent advantage is that it allows the examination of living objects and thus to follow the processes within cells. It was the phase-contrast microscope that made it possible to make mitosis visible and even to film it.

For the procedure itself a special condenser with a ring-shaped mask and an additional "phase-ring" that is fixed within the back focal plane of the objective is needed. The "phase-ring" has two important tasks:

1. It has to achieve an alignment of the brightness of refracted and unrefracted light as the rays that pass through the preparation are weakened in their intensity. In contrast to a conventional image gained with the help of a light microscope, the background of a phase-contrast image is thus dark.
2. The phase difference of most biological preparations is one-quarter of a wavelength or less (*lambda* / 4). The phase-ring is constructed so as to achieve an additional difference of *lambda* / 4. The complete phase difference is thus *lambda* / 2 and crest and trough of refracted and unrefracted rays extinguish each other. A disadvantage of this method is the appearance of light halos around some objects ("halo-effect").

Dark Field Microscopy: This method uses a special condenser with an aperture that big that the light beams that go through it pass by the objective. Only if the object is brought into the center of the light, the light is diffracted, collected by the objective, and used for image formation. Shiny structures are seen in front of a dark background. This method has no very important role in biology, but is impressive with crystals.

FLUORESCENCE MICROSCOPY

Fluorescence microscopy is based on the fact that some molecules emit part of the light absorbed by them as longer waves. A well-known example is the red fluorescence of chlorophyll. Since the Austrian teacher M. Haitinger made his examinations in the thirties of this 20th century it is known that a number of so-called fluorochromes exist, with which microscopic preparations can be stained so that they emit fluorescence indirectly. In the following decades, a whole range of so-called vital dyes were detected or developed, that -used in low concentrations- could mark specific parts of living cells or of tissues. They enabled researchers to follow the solute transport within cells and tissues or to ascertain the pH of special compartments, like for example the vacuole.

Since roughly twenty years fluorescence microscopy is flowering again, on one hand, because the range of fluorochromes became much broader and on the other hand, because completely new approaches like indirect fluorescence could be used successfully. Additionally the construction of microscopes was improved and highly efficient masks were developed.

There are two ways in which fluorescence microscopes can be constructed: as epifluorescence- and as transmission microscopes. The second is the older way of construction. Three components are needed:

1. A strong source of light, that emits mainly short lightwaves. Mercury high pressure lights have proven useful.

2. The first barrier filter: This filter helps to shut off all radiation other than the one that activates the specific dye. It is placed behind the light source within the light cone. Besides it is advantageous to work with a dark field condenser.

3. Second barrier filter: This filter is brought into the light cone between objective and eyepiece. It lets through only long wavelengths, that are caused by emission of the preparation (so-called "secondary radiation").

In the last years the epifluorescence microscope has replaced the transmission fluorescence microscope more and more. But the transmission fluorescence microscope is still better suited to only weakly magnifying objectives (2.5 x, 6.3 x). With the epifluorescence the objective is also a condenser and the stronger it is the more intensive radiation can be used. The heart of epifluorescence is a construction within the light cone between objective and eyepiece that serves to feed activating radiation into the system and is constructed of first barrier filter, beam-splitting mirror and second barrier filter.

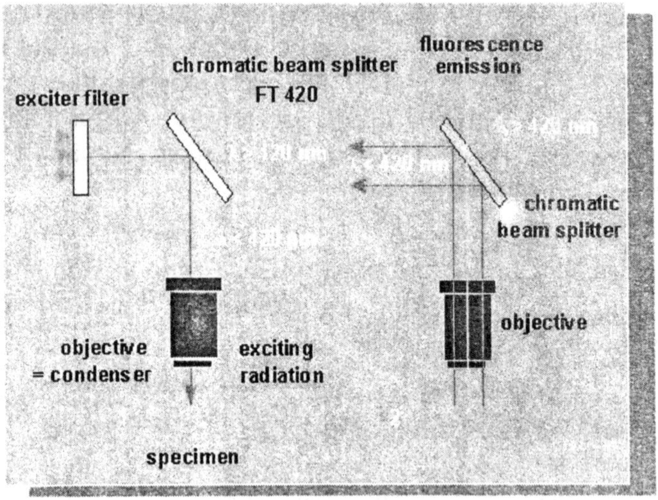

Figure 2. Excitation and fluorescence with chromatic beam splitters. Similar to the interference filters these are specially coated mirrors used under 45° to the illuminating beam. They reflect certain spectral ranges, while others are completely transmitted. The separating line between reflection and transmission may be set at any point of the spectrum. 1. *Exciting radiation,* 2. *Fluorescence emission.*

POLARISATION MICROSCOPY

Light waves do usually radiate into all directions. Suitable polarisation masks allow the elimination of certain planes of polarisation, whereby linear polarised light is gained. It can be completely erased by a second polarisation mask that is placed at a right angle to the first one. Such polarisation masks that can rotate around their own axis can be placed in the light cone of a microscope. One beneath the condenser and the other above the objective.

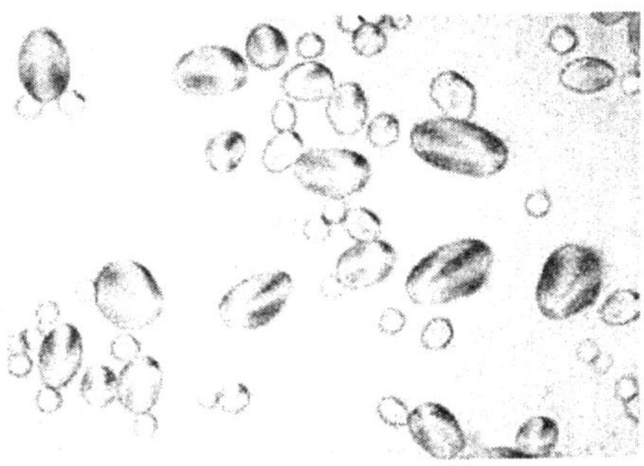

Figure 3. Potato starch in polarised light

The use of a thus constructed polarisation microscope is only sensible, if preparations with polarising properties are examined. This is the case, if the preparation consists of units that are oriented in a certain way like molecules or atoms (double refracting crystals). This type of microscope is thus used mainly in mineralogy since crystals are *per definitionem* built in a regular pattern. By the use of polarised light crystal axis and three-dimensional can pattern be ascertained exactly. The practical application in biology is somewhat limited, but the structure of starch grains for example or the orientation of cellulose fibrils within the vegetable cell wall or the orientation of rod-shaped viruses can be determined.

Based on polarisation microscopy the French physicist

microscope in Paris during the middle of the 1950s. For this type of microscopy two additional Wollaston-prisms are needed. A Wollaston-prism is composed of two cemented calcium fluoride wedges. A polarised ray of light is split into two beams at the cemented plane that are vertical towards each other . The first Wollaston-prism is used in the front focal plane of the condenser, the second in the back focal length of the objective. The object is thus passed by twc beams that are vertical to each other. The phase of the beams is shifted according to the thickness and the refraction characteristics of the object. Optimal interference contrast is reached at the rims of the object, where the phase of both beams is shifted differently. It is important, however, how the object is orientated. A rotating microscope stage is used to analyse the object in all its orientations. The second Wollaston-prism reunites the rays. To achieve interference the planes of oscillation have to coincide. This is done by the polarisation mask above the objective. An interference contrast image looks like a three-dimensional relief. An inexperienced microscoper could thereby be lead to the misapprehension of a three-dimensional structure of the preparation. This is not the case. The tree-dimensional appearance stems from the fact that the differences in the density of the preparation are transformed into differences in height in the image. In contrast to phase-contrast microscopy even relatively thick preparations can be analysed.

Besides the light-dark-contrasts, that can be enhanced or diminished by the rotation of the polarisator or by the adjustment of the second Wollaston-prism, colour contrasts can be achieved by the use of a *lambda*/4 mask.

MICROPHOTOGRAPHY

The development of microscopy created a need for the documentation of the images. The classical approach that is practised by every biology student is drawing. A help was the construction of a drawing device invented by E. ABBÉ, that is placed upon the eyepiece and projects the image via prism and mirror onto a drawing pad. An alternative came into being with microphotography, that has the disadvantage of a very small depth of field and captures consequently only single planes of the object. This is the reason why it is until today viewed rather critically by some biologists. But it has nevertheless been established as an independent method.

Numerous images on our web pages show that an exact image of what can be seen is given by a photo. If it is necessary, several images of different planes of the object can be taken. The important advantages compared to drawings are the speed of documentation and the exclusion of subjective perceptions (freedom of the artist), whereby a lot of damage can be caused. One and the same object can be studied in parallel with with different microscopic methods, corresponding pictures can be compared in order to determine their orientation. The film material of today brings astonishing results even at the very poor light conditions of fluorescence microscopy.

The possibility of cinematography has already been mentioned shortly. Many documentaries of cellular motion, of cells, their coupling etc. exist. Most of the documentaries produced in Germany were done in co-operation with the Institut für den Wissenschaftlichen Film in Göttingen. It is possible for other institutes and other scientific institutions to borrow them.

Video cameras, image intensification and computer-controlled image-storing and -interpretation do also belong to modern microscopy. They have first been used at the beginning of the fifties, but the important developments started during the last years of the seventies. Two processes especially that have been developed by R. D. Allen, N. Stromgren-Allen and J. F. Travis have to be mentioned:

— AVEC-DIC (*Allen video-enhanced contrast-differential interference contrast*) and

— AVEC-POL (*Allen video-enhanced contrast-polarisation microscopy*)

Both methods are based on the fact that certain video cameras can process differences in brightness on a completely different scale than the human eye or a photographic film. Depending on the study or the object can different adjustments of the camera be chosen. E. ABBÉ defined the resolution limit of the light microscope as that distance between two adjoining points, at which they can only just be perceived as different unities. The limiting factor is the wavelength of the used light. This resolution limit does not mean that smaller structures stay invisible. Fluorochromisation, for example, allows single molecular complexes to be seen, if the distance between molecules of the same kind is above the resolution limit of the microscope.

Instead of fluorochromisation contrast-enhancement by AVEC-DIC can be used and although the contrast produced by microtubuli is not big enough to be perceived by the human eye it can be recorded and identified by a camera.

Image-storing and-interpretation allow the emphasizing of movable particles. The appropriate computer program subtracts the content of one image from that recorded seconds or minutes afterwards of the same spot. All immovable structures vanish and only the moveable ones are depicted.

Cameras with image intensification are used in fluorescence microscopy to document fluorochromes with a fluorescence that is far below that perceivable by the human eye. Again it is possible to use the already mentioned method of vital staining, because far more specific sondes are known today. Since their concentration is smaller than that used in the forties or fifties the cells are damaged less.

While in the nowadays common fluorescence microscopy photons (h*ny*) are emitted and recorded after the illumination of the object, a photoelectron microscope allows the tracing of secondary electrons, that are set free after exposure to UV radiation. This microscope has electronic optics and the resolution is higher than that of fluorescence microscopes. The technique has only recently been developed, but first results with biological objects exist already. They suggest that the method will be widespread within short time.

ELECTRON MICROSCOPE

In 1924 L. de Broglie discovered the wave-character of electron rays thus giving the prerequisite for the construction of the electron microscope.

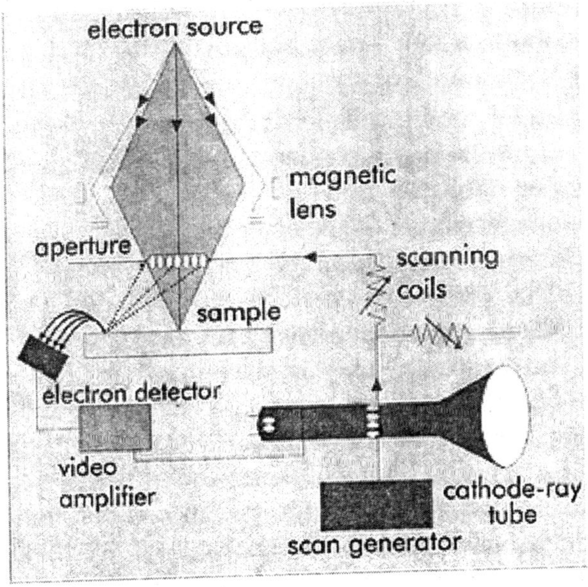

Figure 4. Electron microscope

The prototype was built by M. Knoll and E. Ruska. One of the first biological objects depicted was the tobacco mosaic virus (TMV). The first picture of a cell was published in 1945 by K. R. Porter, A. Claude and E. F. Fullam.

Transmission Electron Microscope (TEM)

The conventional electron microscopy is nowadays called TEM (transmission electron microscopy). We will therefore start with its construction. The ray of electrons is produced by a pin-shaped cathode heated up by current. The electrons are vacuumed up by a high voltage at the anode. The acceleration voltage is between 50 and 150 kV. The higher it is, the shorter are the electron waves and the higher is the power of resolution. But this factor is hardly ever limiting. The power of resolution of electron microscopy is usually restrained by the quality of the lens-systems and especially by the technique with which the preparation has been achieved. Modern gadgets have powers of resolution that range from 0,2 - 0,3 nm. The useful resolution is therefore around 300,000 x.

The accelerated ray of electrons passes a drill-hole at the bottom of the anode. Its following way is analogous to that of a ray of light in a light microscope. The lens-systems consist of electronic coils generating an electromagnetic field. The ray is first focused by a condenser. It then passes through the object, where it is partially deflected. The degree of deflection depends on the electron density of the object. The greater the mass of the atoms, the greater is the degree of deflection. Biological objects have only weak contrasts since they consist mainly of atoms with low atomic numbers (C, H, N, O). Consequently it is necessary to treat the preparations with special contrast enhancing chemicals (heavy metals) to get at least some contrast. Additionally they are not to be thicker than 100 nm, because the temperature is raising due to electron absorption. This again can lead to destruction of the preparation. It is generally impossible to examine living objects.

After passing the object the scattered electrons are collected by an objective. Thereby an image is formed, that is subsequently enlarged by an additional lens-system (called projective with electron microscopes). The thus formed image is made visible on a fluorescent screen or it is documented on photographic material. Photos taken with electron microscopes are always black and white. The degree of darkness corresponds to the electron density of the candled preparation.

Scanning Electron Microscope (SEM)

The path of the electron beam within the scanning electron microscope differs from that of the TEM. The technology used is based on television techniques. The method is suitable for the depiction of preparations with conductive surfaces. Biological objects have thus to be made conductive by coating with a thin layer of heavy metal. The power of resolution is normally smaller than in transmission electron microscopes, but the depth of focus is several orders of magnitude greater. Scanning electron microscopy is therefore also well-suited for very low magnifications. Numerous examples will be given in the following.

The surface of the object is scanned with the electron beam point by point whereby secondary electrons are set free. The intensity of this secondary radiation is dependent on the angle of inclination of the object's surface. The secondary electrons are collected by a detector that sits at an angle at the side above the object. The signal is then enhanced electronically. The magnification can be chosen smoothly and the image appears a little later on a viewing screen.

The properties of the light microscope as opposed to that of transmission and scanning electron microscopes are collected in a table.

Finally some outlines of new and further developments are given.

1. *The high voltage electron microscope:* it operates with an accelerating voltage of 700 - 3000 kV. Its power of resolution is greater, the preparation can be thicker, the strain on the preparation is smaller. But the enormous technical expenditure is disadvantageous. Only few gadgets exist. New results concerning botany have not been gained.
2. *The scanning transmission electron microscope (STEM):* In this development of the SEM do the electrons pass through the preparation and the secondary radiation thus generated is used for image formation. Here, too, the expenditure is large, but it is still worthwhile, since large molecules like nucleic acids or proteins or molecular complexes like viruses can be depicted much better and gentler than with the TEM. No news for botany, though.

The interpretation of images gained with electron microscopy is increasingly done with computerised interpretation programs. But they are usually only suitable for the reconstruction of regularly recurring patterns and these, again, are found more often on a molecular level than on a cellular one.

Preparation Techniques

Biological objects can only be examined with the electron microscope after long and careful preparations. The danger of artefacts is readily given. D. W. Fawcett wrote in 1964:

> "It has to be agreed without reservations that we have no objective criteria to judge the good maintenance of the examined structures. Maybe it is more a belief than a proven fact among morphologists that an image that is clear, continuous, well-ordered, detailed and generally aesthetic is more likely to represent reality than one that is rough, unordered and blurred. But to choose any other criteria as a basic principle would mean encouragement of less carefulness and technical bungling."

Years of research and results achieved with different methods finally guaranteed the reliability of the methods and results. Biological preparations—even single cells—are usually too big and too thick to be used as a whole. Normally cross-sections have to be prepared. Sections require the following steps:

1. Fixation of the material, usually with glutaraldehyde (covalent cross-linking of protein molecules) and osmium tetroxide (binds to and stabilises membranes). Dehydration of the specimen.
2. Permeation with a monomeric resin that polymerises to form a solid block of plastic. Without the thus strengthened structures the specimen would collapse in the vacuum of the microscope.
3. Cutting of the specimen: needed is an ultramicrotome that can produce sections of about 15 - 100 nm thickness (about 1/200 of the thickness of a cell). Ultramicrotomes do normally have fine glass knifes. Edges of break of glass are sharper than metal knifes, but they do not last very long. Diamond knifes are an alternative. They live longer, but are also much more expensive.
4. The thin sections are placed on a small circular metal grid that is coated with a coal-strengthened plastic (formvar) for viewing in the microscope.
5. Two different methods for contrast enhancement exist: coating and impregnation with heavy metal ions. Coating is achieved by placing the specimen in a vacuum and exposing it to a cloud of metal dust (platinium, platinium/coal, gold, vanadium, chrome, lead, etc.). The cloud is produced by the heating of a metal filament that is placed at a defined distance and angle from the specimen. The relief-like surface of the specimen helps the formation of irregular structured metal coats on the specimen (relief contrast). The

imprint of a specimen can also be taken. Therefore slightly thicker metal coats are prepared that are taken off the specimen before viewing them in the microscope.

When contrast is achieved with heavy metal ions, then the preparation is impregnated with uranyl acetate- or lead citrate solutions. The salts are absorbed by the specimen with different strength, so that differently labeled structures are viewed in the microscope later on. It is spoken of positive staining, if a special structure has absorbed the ions and of negative staining, if the metal ions (phosphoric tungsten acid, uranyl acetate, uranyl formiate and others) accumulate around the actual structure. Negative staining is normally used to make macromolecules and molecular complexes (ribosomes, viruses) visible. Usually a special chemical has to be added that prevents the molecules from getting lumpy.

FREEZE-ETCHING AND FREEZE-FRACTURE

Freeze methods (like freeze-drying) offer an alternative to chemical fixation with often better preservation of the specimen's structures. Freeze-fracture and freeze-etching are well-suited for small specimen: cells and subcellular structures. The specimen is frozen and the frozen material is "broken" with the aid of a glass knife. The preparation splinters and the edges of breaking run along the membrane or between two half membranes.

Water is subsequently removed *via* freeze drying. It sublimates, i.e. it transforms out of its solid phase directly into the gas phase. This process etches the surface of the preparation. It is then coated and the metal coat is taken off and viewed in the microscope.

During the late 1970s, J. E. Heuser developed the quick-freeze, deep-etch preparation of samples that avoids the crystallisation of water. The (very small) specimens are frozen quickly in liquid nitrogen transferring the water into a glass-like state. This leads to a much better preservation of the sample. The method is suitable for the depiction of large molecules and molecular complexes. Here, too, an imprint is gained by coating. Contrary to negative staining a three-dimensional image of the specimen can be formed.

REFERENCES

Allen, R. D., N. Strömgren-Allen, and J. F. Travis, "Video-enhanced contrast, differential interference contrast (AVEC-DIC) microscopy", *Cell Motility*, 1981.

Allen, R. D., J. L. Travis, N. Strömgren-Allen, and H. Yilmaz, "Video-enhanced contrast polarization (AVEC-POL) microscopy", *Cell Motility*, 1981.

Hearle, J. W. S., J. T. Sparrow, and P. M. Cross (eds.), *The use of scanning electron microscope*, Oxford: Pergamon Press, 1972

Heywood, V. H. (ed.), *Scanning electron microscopy*, New York, London: Academic Press 1971.

Tanaka, K., "Scanning electron microscopy of intracellular structures", *Intern. Rev. Cytology*, 1980.

2

General Features of Plant Cells

Cells come in a large variety of forms. There is no such thing as a typical plant cell. The typical plant cell given in some textbooks displays a compendium of features of different cell types. All cells do, nevertheless, share a number of features, while some are typical only for plant cells. A third group of features distinguishes undifferentiated from differentiated cells.

Every cell is surrounded by a membrane. Both animal and plant cells contain a nucleus that is missing in bacteria and blue green algae. It is thus distinguished between eucaryotes (organisms that have a nucleus) and procaryotes (organisms without a nucleus). Contrary to animal cells plant cells are almost always surround by a cell wall and many of them contain a special group of organelles: the plastids, of which the chloroplasts are the most striking organelles. Cell walls can also be found with bacteria and blue green algae, though neither their chemical structure nor their synthesis is the same as that of plants. It does seem to be an analogous feature, i.e. a structure with the same function, which did not derive from a common earlier stage.

The content of a plant cell (the protoplasm) changes in composition in the course of growth and development. Differentiated cells are marked by a voluminous vacuole. Besides the clearly visible structures like cell nucleus and chloroplasts, a lot of other granular particles can be seen through a light microscope, some of which can be dyed selectively. Sometimes, different shapes (long, round, etc.) can be distinguished. The majority of these particles has a size that is at or below the power of resolution of a conventional light microscope and their identification was impossible before the use of the electron microscope. Mitochondria are a typical example.

The organelles may contain inclusions like crystals, fat droplets, starch grains or others, a feature that is especially common in specialized cells. Cells are no static objects. Very often, plasma currents occur that can be detected by the movements of organelles like chloroplasts or different granules. Most of these movements have a direction and sometimes it even looks as if single particles would move along rails.

STRUCTURE OF A PLANT CELL

To depict the structure of a plant cell, an epidermal cell of an onion will be used as an example. The epidermis is the final tissue that covers all organs above ground. The cells of the onion epidermis are common specimens on the first day of a German basic botanical course. Since they contain no chlorophyll, they are actually no "typical" plant cells.

Protoplasm, Cytoplasm and Cytosol

The "living" content of a cell, the protoplasm, is surrounded by a membrane called plasmamembrane or plasmalemma. The protoplasm is usually next to the cell wall, so that the plasmalemma can hardly be seen. To display it, the cells are transferred into a high salt or sugar solution. As a result the protoplasm shrinks and detaches itself from the wall. The process is reversible and is called plasmolysis. This behavior is due to the properties of the membrane and the plasma. It is reviewed in more detail elsewhere. A substance that causes plasmolysis is called plasmolyticum and—depending on its chemical composition (potassium ions or calcium ions, for example)—the protoplasm takes on different shapes. The plasmolyticum has accordingly an influence on the properties of the membrane.

The properties of the plasmamembrane differ from that of the tonoplast. The tonoplast is the membrane that surrounds the vacuole. The difference is especially striking if cells with a colored vacuole content are used. Often the vacuole is criss-crossed by numerous plasma cords. The plasma cannot therefore not simply be viewed as a solution that is influenced by the rules of hydrodynamics alone. Rather, it contains viscous, structure-determining components, whose chemical, physicochemical and structural properties have only been recognised recently and in fragments.

Cytoplasm and Caryoplasm

The nucleus is a rather conspicuous part of nearly every living plant cell. Its structure separated from the rest of the cell by the nuclear envelope, a membrane system that consists of two discrete membranes as can be seen on electromicroscopic images. The nuclear content is called the caryoplasm while that of the rest of the cell is called cytoplasm. But these terms are only valid at certain stages of a cell's life cycle. In the course of cell-division and mitosis, the nuclear envelope disintegrates and the nucleus is replaced by the chromosomes. It makes consequently no sense to speak of caryo- and cytoplasm during these stages.

The nucleus of plant cells is usually of a round or elliptic appearance, sometimes it is also shaped like a spindle. One nucleus per cell is the rule, but cells with two or more nuclei are no rare exception. The cells of certain algae of the genus *Chladophora* have many nuclei, they are polyenergid. The nucleoli that can often be perceived after staining are substructures of the nucleus. They, too, disintegrate during cell division and mitosis and do not reshape before a new nucleus has been formed.

Plastids

Plastids are organelles that occur only in plants. Their most prominent members are the chloroplasts. Others plastids are the colored chromoplasts and the colorless leucoplasts as well as their proplastids. Proplastids are vestigial bodies that are generated during germ cell development due to degeneration of plastids, for example. They may differentiate into complete plastids during the development of the plant embryo. Their ripening into chloroplasts occurs usually only after light exposure.

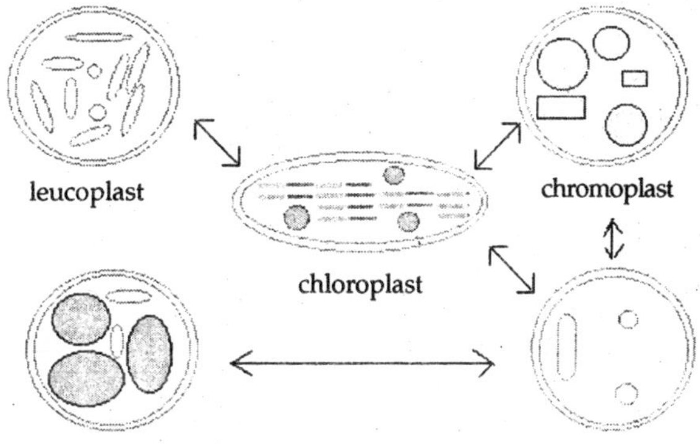

Chloroplasts contain the green plant color chlorophyll. They are the places where photosynthesis takes place. Chloroplasts enable the plants to convert solar energy into chemical energy. Because of this process, plants are called primary producers. The existence of consumers, like most animals, depends on them. Chloroplasts occur in most cell types, but only in organs above ground. They can be especially well observed in tissues consisting of a single layer as in the leaflike structures of some mosses (*Funaria hygrometrica* or Mnium hornum) or in the water plant

Vallisneria). Here they are rather large and of a lens-shaped appearance. During daytime, when the light is diffuse, they occur mainly at the upper and lower surface of the chloroplast. They appear to be round under top view. If exposed to strong light, they gather in parallel to the lateral sides of the cell which gives them an elleptic appearance upon top view.

Chloroplasts are the site of starch production and-storage. Starch can easily be detected with the aid of potassium iodide. The starch-iodine complex is deeply blue-violet. Starch production during photosynthesis can be made visible by placing a mask at a leaf that covers it partially while leaving some places exposed to sunlight. After one day of exposure, the leaf is first bleached to get rid of other pigments and afterwards treated with potassium iodide. An image is gained that is the exact replication of the mask and at the same time represents starch synthesis in the leaf. This experiment has first been done by J. v SACHS, probably the most outstanding plant physiologist of the 20th century. He thought that starch was the primary product of photosynthesis. This assumption proved wrong. It is well-known today that the first products of photosynthesis are monomeric sugars (glucose and others) and that only part of them is used for starch production.

The structures of the chloroplasts of higher cells resemble largely that of mosses. Their average diameter is 4 - 8 mm, an average cells contains 10 through 50. Their chlorophyll is unevenly distributed. At high resolutions chlorophyll-rich and chlorophyll-poor areas can be distinguished. This is due to the inner structure of the chloroplast: it is organised into grana (chlorophyll-rich) and stroma (chlorophyll-poor). Upon stimulation with short-waved light (blue or violet) the chlorophyll emits an intensive red autofluorescence that looks especially impressive in a fluorescence microscope. The differences between grana and stroma become very obvious. The uniformity of the chloroplasts of all higher plants points out that the optimal form has been found rather early in evolution and has not been changed since. This is different with algae. The chloroplasts of green algae (Chlorophyceens) are very varied in shape. Many species have just one chloroplast that covers nearly the whole space of the cell's interior. It is screw-like in *Spirogyra*-species, star-shaped in *Zygnema* and Zygnemopsis and netlike in Oedogonium.

The disc-shaped chloroplast of *Mougeotia* can be viewed either from above or in profile depending on the amount of light used. Its rotation is a well-analysed example of an induced chloroplast movement. The chloroplasts of many species of algae contain often well-visible pyrenoids, structures, that produce and structure starch.

Chromoplasts are red, orange or yellow plastids. The color is usually the result of yellow xantophyll and yellow to red carotinoids. Both compounds do also exist in chloroplasts, but are concealed by chlorophyll. Chlorophyll is broken down much faster than carotinoids as can be observed in the colored leaves in autumn. Fluid transitions between chromo- and chloroplasts exist, just as between chromo- and leucoplasts. Typical chromoplasts cause the orange color of the carrot, the red color of the ripe pimento and tomato as well as the color of numerous flowers. Carotinoids are not very water-soluble and do therefore often crystallise within the chromoplasts. Their crystals can be disc-shaped, needle-like, jagged or sickle-like.

In many cases, flower and leaf colors are caused by the colored content of the vacuole. The color of the vacuole and that of the plastids may lead to a mixed color. The leaves of the copper beech, where the vacuole's content is red and that of the chloroplasts is green are a typical example. The plastids of the red and brown algae are traditionally counted among the chromoplasts although they contain chlorophyll. The green colour is concealed by the red phycoerythrin (*Rhodophyceae*) or the brown fucoxanthin (*Phaeophyta*).

Leucoplasts are common, colorless plastids. They develop from proplastids, but form no homogeneous group of their own. A part of them can differentiate into chloroplasts or chromoplasts at light exposure, while this is not due for others. The guard cells, for example, contain leucoplasts, that are permanently exposed to light without developing into chloroplasts. Leucoplasts do also occur within colorless leaves (variegated leaves) or plant parts. There exists a number of examples which show that they developed from chloroplasts that lost their ability to produce chlorophyll. There are even species, like Neottia, an orchid that cannot produce chlorophyll at all and are thus dependent on a parasitic or saprophytic lifestyle.

A second class of leucoplasts occurs within the non-green tissues of otherwise green plants. It is especially common within roots. Though these leucoplasts are capable to become green, they do usually not since they are not exposed to light. The leucoplasts of the calyptra (a calyptra is any hood or cap of cells protecting a plant part) contain starch and are therefore counted among the amyloplasts (starch-containing leucoplasts). They have, as is explained later, the function of statolithes, that have an important part in the perception of gravity (geotropism).

Starch

Since the plant is able to transport sugars from leaf to root or from leaf to seed and fruit, starch production can also take place in these organs. Different species produce starch grains of different shape. Since the shape of starch grains informs about their origin, they are helpful in the identification of seeds and other starch-containing plant parts. The following numbers show the variations in their diameters. Starch grains from potato tubers have a diameter of 70 - 100 µm, that of the endosperm of wheat 30 - 45 µm and that of corn endosperm 12 - 18 µm. Their shape reflects their development.

The starch molecules are long-stretched and only sparsely branched. They are deposited within the plastids and their development begins at a so-called formation center from where it proceeds radially. Layer follows layer and the thickness of one layer is dependent on the average molecular length. A starch grain is therefore organised like a crystal (semi-crystalline). This can be shown very impressively with a polarisation microscope. Within the moistened specimen can a layering be viewed that is dependent on the water content of the single molecular parts.

A. Model of a starch grain structure. The single lines symbolise starch molecules. They are arranged in a radial pattern.

B. Layering of the starch grains. a. Formation center and layer borders, b. Diagram of the refraction conditions. The ordinate shows the refraction index.

The denser the molecular packaging, the less water is deposited. Layers with less water content refract light stronger than those with much water. After drying of the preparation no layering can be perceived any more. Depending the central or peripheral placing of the formation center, starch grains with either concentric or eccentric layering develop. The starch grains of graminaceous plants (wheat, corn, etc.) are usually concentric, while those of the potato are always eccentric.

Sometimes plastids with two to three formation centres occur in graminaceous plants. This leads to the development of several starch grains. During growth common outer layers may be formed.

Compound starch grains are typical for oat, they are built from a large number of smaller grains.

Starch grains in bean seeds (*Phaseolus vulgaris*) are very big, their shape is round or oval, the spacing of the layers is very regular. Their centers can easily be hollowed out by addition of water displaying radial ruptures in microscopic images. In the sap of *Euphorbia splendens* dumb-bell shaped starch grains can be found.

Crystalline Inclusions

Many plant cells contain crystalline inclusions of different chemical composition and shape. Crystalline aggregations are called druses, bundles of needle-shaped crystals are termed raphids. Scanning electron microscopic images: Top picture: Calcium oxalate druse in the mesophyll cells of an oleander leaf *(Nerium oleander)*. Typical druse shape of dicots. Middle picture: Calcium oxalate needles (raphids) of a vanilla root (Orchidaceae). Typical raphid bundle of monocots. Lowest picture: Silicate bodies of silicate cells in the epidermis of Schizachyrium sanguineum (a gramineaen species of the old world tropics). Characteristic mineralisation of a gramineaen cell.

Cell Wall

Except for very few examples, plant cells are surrounded by a cellulose containing cell wall. It is flexible and distortable during growth, but loses its ability for distortion after growth has stopped, while a limited flexibility remains. Because of these changes, it is distinguished between primary and secondary cell walls. As we will see when talking about electron microscopic pictures of the cell wall, both forms differ mainly in the arrangement of their cellulose microfibrils. While they are unorganised within the amorphous matrix of the primary cell wall, they are organised into several ordered layers that are arranged one on top of the other at right angles in the secondary cell wall. The secondary cell walls of many cells, especially those of vascular tissues, are incrusted with strengthening material. Two important ones are:

— lignin, the ground substance of wood and

— suberin, the ground substance of cork

Their details are reviewed here. In addition, secondary walls contain often phenolic oxidation products that lends them a dark color (red to black with various shades).

Tissues

The branch of science that is committed to the study of tissues is called histology. Though with plants the term plant anatomy is heard just as often. The term tissue stems originally from a misinterpretation. N. GREW coined it in 1682 since he thought that the scaffolding of the cell wall would consist of very fine threads and that the organisation of the plant would resemble the layering of a larger number of Brussels lace on top of one another. Despite this misapprehension the term was used from now on and was even adopted by animal histologists.

From our contemporary point of view tissues are combinations of cells, organs are functional unities of an organism. The basic organs of the phanerogamous plant are leaf, stem and root. Organs consist of different tissues, the leaves for example of dermal tissue, assimilation tissue and vascular tissues. A tissue again may have differently structured cell types. The vascular tissue contains thus the cells of the xylem and those of the phloem. Based upon a suggestion of C. W.v. Nägeli it is distinguished between local regions of cell division, the meristems, and permanent tissues. Meristems are characterised by cell divisions, while this is an exceptional feature in permanent tissues. The latter are as a rule differentiated and often specialised. Depending on their functional properties can they be grouped into the following categories:

— dermal tissues (epidermis, cork, bark)
— ground tissues (parenchyma)
— assimilation tissues (palisade parenchyma, spongy mesophyll)
— collenchyma and sclerenchyma tissues (collenchyma, sclereids, fibers)
— vascular tissues (vessel elements, xylem, phloem)

Probably the most important principle of vegetable differentiation is the polarity that is developed during embryogenesis (ontogenesis) at the first division of the fertilised egg. It determines the main axis of the vegetation body irreversibly. After that, shoot and root develop independently. This polarity is also known as root-shoot-polarity.

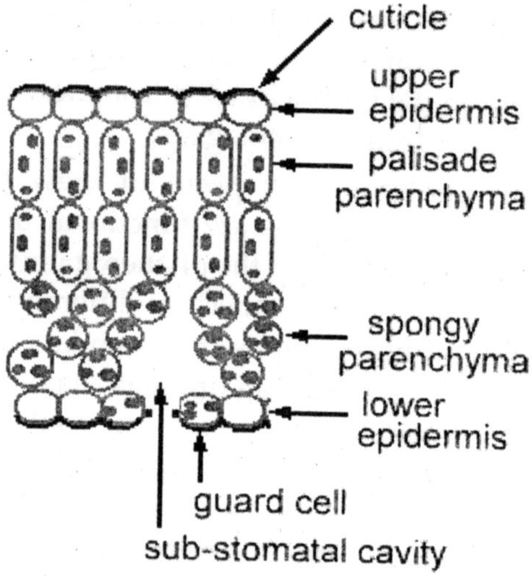

Figure 1

Cell divisions that take place in a plane that is perpendicular to the surface of the next surface of the organ are called anticlinal. Those that take place in parallel to it are called periclinal. Often the cells divide unequally and as the result are two cells of different sizes. The rule of thumb says that the smaller one remains in the already existing physiological state, while the bigger one differentiates or specialises into a certain direction. Exceptions exist. In guard cell development, for example, the smaller cell differentiates stronger than the bigger one.

One of the most striking features of unequally differentiated cells is their uneven enlargement. Some cells divide without noticeable increase in volume, while others stop dividing and grow considerably. Vegetable growth is thus caused mainly by an increase in the volume of single cells. Since the enlargement of every cell is restricted by the resistance of neighboring ones and since the cells are cemented to each other rather strongly by the middle lamina, considerable tensions are inevitable. A cell that enlarges evenly into all directions results in a sphere. Neighboring cells have different sizes due to both asynchrony of cell division and different individual ontogenesis, so that the walls between neighboring cells differ in size, too. As a result, the shape of a cell that enlarges evenly into all directions looks more like a many-sided body, a so-called isodiametric cell or polyhedral, than like a sphere.

Elongation of cells is often directed. It occurs usually in parallel to the axis of the organ in question. The resulting cell shape is that of a spindle. It is also called prosenchymatous. If all cells of a certain area are elongating (like those of the apical meristems of roots and shoots), then the respective organ elongates evenly, while the growth of cells at only one side of the organ will lead to bending.

The mentioned tensions and altered plant shapes, that are caused by uneven cell divisions and differences in volume enlargement of single cells show that the cormus of the plant cannot be described by simple geometric relations.

The irregularities of cell divisions (different division activities of single cells or cell groups, unequal divisions, etc.), the differences in the ability to elongate and the way of specialisation are therefore the causes of the histological variability and the formation of a specific shape of the plant (morphogenesis).

Although every step itself seems like a deviation of the normal, all steps are controlled by the plant's genome. Growth and differentiation are thus exactly matched and co-ordinated in a way that gives the developing plant the form that is specific for the respective species. There is accordingly a flow of information between the cells of an organism that helps regulating their activity. This is most impressive in the development of symmetric shapes at both cellular and organ level. But how is the control of growth and the co-ordination of differentiation achieved?

Meristems

Cormophytes produce new cells throughout their life and form new organs at regular intervals. All cells of an embryo display the same activity of division, but as soon as a certain state of differentiation and thus a certain size has been reached, the production of new cells is restricted to special tissues, the meristems.

Typical meristems are found at the tips of all stems and roots. They are called apical meristems. Those of the shoot are usually protected by involucral leaves and the whole complex forms a bud. The apical meristems are the cause of tip growth that is one of the most striking features of vegetative growth. This contrasts with animal growth, that is an allometric process, i.e. all parts of the body grow in a weay that maintains the body's overall proportions. As soon as an animal has reached its final size, growth ceases. This, however, does not mean that no more cell divisions are taking place. Cell divisions occur all the time, especially in epithelia and in blood stem cells. But this division activity has always the function of replacing old or damaged cells. No such repair occurs in plants. If a leaf or flower or any other part is damaged, it is neither replaced nor repaired. Instead, new organs grow, and this again does not happen in animals.

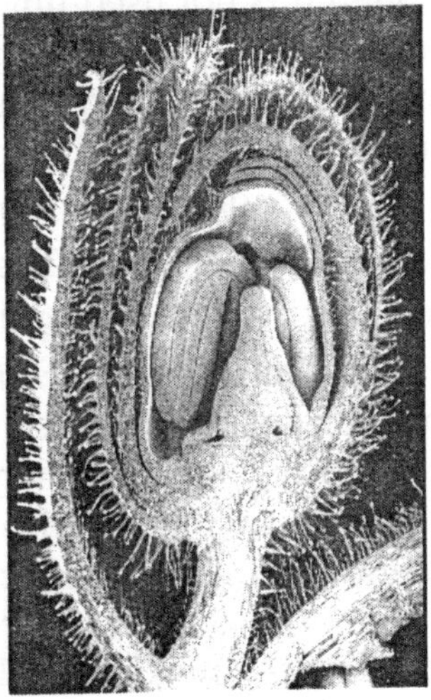

Figure 2. Immature Flower Bud—Snapdragon (Antirrhinum majus)

The cells of have a pability to divide. But division is controlled. Many plants, for example, follow the principle of apical dominance, where the activity of the lateral meristems is suppressed while the tip is growing. In this case, information is exchanged between the two tissues in question. If the tip of the main axis is cut off, suppression is abolished and the lateral buds start to sprout.

Many plants like gymnosperms and dicots have extensive lateral meristems (the cambium, the vascular cambium, and the cork cambium) that give rise to growth in girth, also called secondary growth. Their activity may fluctuate in the course of the year and is reflected by annual rings.

Other plants, like monocots, have intercalary meristems. They are actively growing meristems that differ clearly from apical meristems and are located between more or less differentiated tissues usually at the base of each internode. The formation of secondary meristems shows that differentiated cells can reverse their state and go back to a meristematic existence. Their ability to divide is thus not lost, though it is not used unless an adjustment to changed circumstances requires it. The ability of indeterminate growth is widely exploited by the use of cuttings in horticulture. It is also displayed in the formation of roots at the cutting sites of begonia leaves, for example.

But what exactly is a meristem? The following three examples will help to answer this question.

The marine brown alga *Dictyota dichotoma* consists of a flat thallus with a thickness of three cell layers. The large cell at its apical tip divides periclinal, while the subapical progenitor cells divide in an anticlinal manner. The progenitor cells themselves display alternating peri- and anticlinal cell division. The thallus enlarges thus both in length and in width. At regular intervalls, the apical cell divides in an anticlinal manner, too, resulting in two apical cells and a subsequent forking of the thallus. Although this example explains the way a division scheme works, it does not explain where the meristematic cells end and where the differentiated ones begin. The classical scheme of a cormophyte given by Sachs says that no clear demarcation exists. Instead, there is a gradient between meristematic and differentiated cells. The meristematic properties are thus not lost during one division, but they decline gradually.

The second example shows how division takes place in some liverworts, in simple cryptogams like horsetail and in many ferns. Contrary to that of many algae, their cormus normally resembles a three-dimensional body with a single tetraedical initial cell at the tip of the axis. It produces progenitor cells by regularly alternating the planes of division. These progenitor cells are given off towards the base. The progenitor cells may again divide in both anti- and periclinal manners.

The tip of phanerogams, my third example, is formed by a whole group of cells. They are not specialised and therefore called vegetation zone. The apical meristem is organised into several cell layers. In angiosperms and in some gymnosperms, it has to be distinguished between the outer tunica and the central corpus. The meristematic cells divide mostly in an anticlinal manner, while those of the corpus display both planes of division: anti- and periclinal. Some parts of the corpus are dominated by a certain plane of division, thereby determining the first step of differentiation. Usually, the outer layer of the tunica produces the epidermis. The inner tissues of the plant stem from the corpus, the tunica or both.

These schemes are useful guidelines for the understanding of the plant's morphogenesis, since they point out the importance of the information that is contained in the cell's position within the tissue. A cell will only develop into a certain direction, if its position within the tissue is "right". If a cell is transferred to a new position, it will adopt new functions and thus dedifferentiate. At the beginning of the 20th century, G. HAberland put forward the sentence of the totipotency of plant cells. It states, that each cell of a plant keeps the ability to develop a complete plant. This is true for most, but not all plant cells. And still: growth through division is in fact characteristic for the meristematic state of a cell, but it is by no means restricted to it. A meristem does not only include the initial cells and their immediate progenitors, but also some parts of the shoot.

The situation is similar in root meristems, although their anatomy is different. The shoot apex is protected by the involucral leaves. In roots, this function is exerted by the root cap. Its cells are produced by a specialised root cap meristem and are progressively replaced towards the tip.

These cells have to have a high turn-over, since the outmost cell layer is easily damaged by soil particles as a result of the growth movement of the root.

Procambium and Lateral Meristems

It is normally distinguished between primary and secondary growth of the shoot. The first is the phase of plant development that gives rise to new organs and to the basic plant form, the latter brings about growth in girth and the formation of new vascular tissues.

The procambium is the meristematic tissue that produces the primary vascular tissues: xylem and phloem. It develops directly beneath the growing tip next to new leaf primordia. The development of new leaves is thus tightly connected with that of new vascular tissues. The vascular tissues of leaves are also called leaf veins. The cells of the procambium are generally combined in cords. They form an extension of the vascular tissues into the growing tip and do thus provide the connection of the newly developed organ with the conductive systems of the plant. The cells of the procambium are elongated and become even more so in the course of their development. The volume of their vacuoles increases considerably, lending them a lighter and more transparent appearance than their neighboring cells. This is decisive for their differentiation into xylem or phloem cells.

Some of the meristematic cells in plants with secondary growth keep their meristematic state and become cells of the cambium.

What are the causes for the development of the procambium? Two models exist:

1. They are to be found in the cell-cell interactions that take place between the cells of the bud, that determine the pattern of the leaf positions and the positions of xylem and phloem.

2. Induction is achieved by fully differentiated tissues. The necessary information is conducted to the bud via already differentiated cells of the procambium.

The first model was confirmed experimentally, since formation and further development of the procambium are undisturbed in isolated bud meristems. The determination of cells takes place before any morphological change can be seen. Differentiation describes the (usually irreversible) changes of structures and functions that result in a certain specialisation, while determination means the (usually irreversible) triggering of processes that lead to these changes.

The cambium is the prototype of a lateral meristem. It is mono- or multilayered depending on its origin and forms normally a continuous cell layer of tubular shape that is located at the periphery of the shoot or root. It separates xylem from phloem, if present. The cambium develops from the procambium next to the vessels, ensuring a continuity of the meristematic state.

The cambium between the vessels derived from already differentiated, parenchymatous cells. The first type is also called fascicular (within the vascular bundle) cambium, while the latter one is termed interfascicular (between the vascular bundles). The interfascicular cambium is a secondary meristem. A few monocots (*Dracaena, Yucca, Aloe* and others) have again another type of cambium, the extrafascicular cambium.

The cells of the cambium are often termed initials, since they initiate the formation of specialised progenitor cells after division. Two cell types occur within the cambium:

1. *Fusiform initials.* They are the mother cells of all xylem and phloem elements, as well as of all other cells that are oriented parallel to the organ's axis. They are flat, elongated with pointed ends and highly vacuolate. It is their spindle-like shape that caused their name.

2. *Ray initials are nearly isodiametric, small cells that occur often in groups.* They develop from fusiform initials or their progenitors. They produce the radially orientated rays in wooden plants (transversal elements).

I. W. Bailey, who did the basic work on the organisation of the cambium in the 20th of this century, gained the following data when comparing a one year old and a sixty years old stem of *Pinus strobus.*

Table 1. Differences in cambium girth and in number of initials between a one year old and a sixty years old stem of Pinus strobus

	Age 1 year	60 years
radius of the tube	2 mm	20 cm
girth of the cambium	12,56 mm	1,25 m
average length of fusiform initials	870 µm	4000 µm
number of fusiform initials in cross-section of stem	724	23,100
number of ray initials in cross-section of stem	70	8796

His data show that both length and number of fusiform initials increases with the age of the stem. The increase in number and the widening (dilatation) of the cambium tube is caused by the widening of the central xylem cylinder, that is produced by the cambium itself.

Secondary vascular elements are given off into opposite directions by periclinal (tangential) division of fusiform cells. The developing xylem elements grow towards the inner cylinder, that of the phloem towards the outer of the cylinder. The progenitors of the initials are organised in radial rows that make it easy to trace back their descent. The increase in girth is compensated by the anticlinal divisions occurring at regular intervals. The activity of the cambium is temperature-dependent: it causes the annual rings. Their thickness, i.e. the activity of the cambium, is determined by extern factors like temperature, day length, soil humidity, temperature and others. Besides its function in the production of vascular tissues, the cambium has an additional capacity for the healing of wounds.

The cork cambium (or phellogen) is a secondary lateral meristem that serves to produce the secondary outer surface, the bark, that replaces the epidermis. It is without exception given off towards the outward direction. Often, though not always, the cork cambium produces cells towards the inner of the stem, that form the phelloderm.

The cork cambium has a rather simple structure, if compared to the cambium. In cross-section, the cells have a right-angled shape, they are flat in radial and tangential sections. Their plasma is highly vacuolated and may contain chloroplasts and tannic acids.

The cork cambium stems from epidermal cells or/and from cells of the underlying parenchyma. It is distinguished between primary and secondary cork cambium taking into account the fact that it can be assembled within the stem several times during a plant's life. The cork cambium is monolayered during the first year in some species and becomes multilayered later on. It may be active for several years, sometimes even for the whole life or for just one year. Its activity is, just as that of the cambium, influenced by extern factors.

REFERENCES

Bierhorst, D. W., *Morphology of vascular plants*, New York: The Macmillan Comp, 1971

Esau, K., *Anatomy of seed plants*, New York: J. Wiley and Sons, Inc., 1960

Fahn, A., *Plant anatomy*, Oxford: Pergamon Press, 1967

Yamaguchi, Y., R. Nagai, "Motile apparatus in Vallisneria leaf cells", I. Organization of microfilaments, *Journal of Cell Science*, 1981.

3

Dermal and Assimilation Tissues

The epidermis is the main dermal tissue of primary plant organs above ground. It covers the shoot, leaves, flowers, fruits, and seeds and serves several functions. Among them are protection from water loss, against physical and chemical influences and from feeding by animals, as well as participation in gas exchange and secretion. Special cells or devices for some of these functions developed during evolution. Light and mechanical stimuli are perceived by specific receptors, the perceptors.

The stoma complexes are scattered across the whole epidermis and act as sites of gas exchange between leaf and atmosphere. Stoma complexes consist of several cells. Two guard cells that enclose the actual stoma, a pore in the epidermis, control its state of opening. They are supported by two or more subsidiary cells. The cuticle, a structure that is secreted by the epidermal cells and covers them like an uninterrupted film helps to prevent water loss. This is also a purpose of the trichomes, epidermal hairs. But not the only one. Trichomes function as devices against animals, as glands, as weapons or water vesicles, too.

Another dermal tissue is the rhizodermis, which covers all underground plant parts. Its main feature are the root hairs. They enhance the plant's surface and consequently also its ability to absorb enormously. Cells of the rhizodermis secrete a mucilage, that generates a rhizosphere, whereby absorption is further contrived.

Older shoots and roots have normally gone through secondary growth. Their primary dermal tissues are replaced by a secondary dermal tissue, the bark. The bark is produced by the phellogen, a secondary cambium, and is interspersed with groups of cells called lenticels. The lenticels have the same function as the stoma complexes in leaves: they are sites of gas exchange.

Ground tissues make up the bulk of all tissues. All specialised cells are embedded in them. The cells of the ground tissue are the parenchyma cells. Parenchyma cells have assimilation, storage- and wound-healing functions. Parenchyma tissues are the mesophyll, an important assimilation tissue, the palisade parenchyma and the spongy parenchyma of leaves. Further leaf structures are the vascular bundles and their bundle sheats. A specialised way of carbon dioxide fixation is performed by a wreath of cells that surrounds the bundle sheats of C4 plants, the Kranz anatomy. The development during ontogenesis determines the leaf's shape. Its growth takes place in wave-like growth movements.

CELL TYPES OF THE EPIDERMIS

There exist two techniques for the depiction of the structural organisation of epidermal surfaces. One is scanning electron microscopy, the other the microscopic analysis of imprints in synthetic material. Both techniques are demonstrated by a number of examples.

The epidermis has numerous functions:

— protection against various chemical and physical influences, against being fed upon by animals and against infestation by parasites

— protection of the plant against desiccation

— participation in gas exchange, in secretion of metabolic compounds and in absorption of water

— site of receptors for light and mechanical stimuli that help to transform signals from the surrounding to the plant

The epidermis has accordingly a number of differentiated cell types to serve the various functions. Variations typical for certain species and different organisations of the epidermis in the miscellaneous plant organs add to the number of different cells. Three main types exist:

1. the basic epidermis cell
2. the cells of the stoma complexes and
3. the trichomes (gr.: trichoma = hair), epidermal attachments of varying shape, structure and function

The basic epidermis cells, i.e. the least specialised cells constitute the largest group of dermal cells. They seem either polygonal or elongated in top view. Their walls are often wavy or sinuate. It is unknown, what induces this shape during development, since the explanations given by the existing hypotheses seem insufficient. Elongated epidermis cells can be found at organs or parts of organs that are elongated themselves, like stems, leaf petioles, leaf veins or leaves of monocots. The epidermises of the leaf's upper- and undersurface may have different structures. The shape of the cells, the thickness of the walls as well as the distribution and number of specialised cells (guard cells and trichomes) per area may all vary. Wide varieties of different cell shapes may even exist in species of a single family, e. g. in the Crassulaceae family

The wall of epidermal cells that constitute the leaf's surface is often thicker than the other walls. This can be particularly well observed with the epidermis of conifer needles and that of xerophytes (plants living in dry habitats). Aquatic plants have usually thin walls. The wall of many seeds becomes stronger during ripening and may fill nearly all of the cell's lumen so that the protoplast is driven out and degenerates. The basic epidermal cells of most species contain no chloroplasts. Some ferns and several aquatic or shade plants are exceptions.

The epidermis is more often than not built from a single cell layer, though multi-layered, water-storing epidermises that evolved from initially single-layered tissues by periclinal division have been shown among the species of several families (Moraceae: most Ficus-species, Piperaceae: *Peperonia*, Begoniaceae, Malvaceae and others). Epidermis cells secrete a cuticle, that covers all epidermal surfaces like an uninterrupted film. It may either be smooth or structured by bulges, rods, filaments, folds, or furrows.

The folding of the cuticle is nevertheless not always caused by cuticular rods or filaments. There are cases where it is merely the expression of the bulging of the cell walls below. It may be necessary to analyse cross-sections of the cuticle and the underlying epidermal cells to find the actual cause.

In some cases, like that of the tomato fruit, the cuticle is pigmented with carotinoids. Often, additional waxes, oils, resins, salt crystals and (hydrophilic) mucilage are excreted. The latter is especially common in developing seeds. The excretion of wax inhibits the moistening of the leaves more than the cuticle itself. These waxes do often have an intrinsic structuring. A thick layer of wax lends a whitish appearance to the plant surface. It has two effects: diminution of water-loss and enhancement of reflection of the sunlight, thus protecting the plant from too much heat.

Stomata and Guard Cells

The stomata accomplish an important function of the epidermis. Their complete functional unit is the stoma complex. It is composed of two chloroplast-containing guard cells that enclose a pore (the actual stoma), as well as of two to four subsidiary (or accessory) cells devoid of chloroplasts. Cross-sections show that the guard cells have walls of different thicknesses. They sit at the top of an intercellular space, that is in communication with other intercellular spaces of the respective tissue. The stoma can be opened or closed according to demand. In this way, the guard cells regulate both transpiration and CO_2- uptake. Water-content and CO_2- concentration within the plant tissue act as regulators for the state of opening. The guard cells control the size of the stoma by changing its shape: more about the mechanism.

Stomata occur at all plant surfaces above ground. Their number approaches 100-300 per mm^2, but there exists a wide variation in number and distribution. They are lacking in roots, in the epidermis of the chlorophyll-free shoot of the parasitic species *Monotropa hypopitys* and *Neottia nidus-avis* as well as in some submerse living aquatic plants, while they are normal in others. They do, too, happen to be present in some colored or white petals, though they have often lost their function here. The parallely veined leaves of most monocots, some dicots and the needles of conifers have stomata that are grouped in parallel rows.

Stomata are produced by the guard cell mother cells, that again are laid out within the epidermis in regular intervals. The initial of the stoma complex is the smaller of the two progenitor cells. It contains more plasma. Both guard cells stem from it and are formed by unequal division, while the production of the subsidiary or accessory cells may occur in many different ways. The development of a leaf's stoma happens mostly asynchronously. It follows the successive differentiation of the single tissues in leaves with parallel venation. Differentiation proceeds in a basipetal manner, i.e. it starts at the tip of the leaf and continues towards its base.

Netted leaves have stomata of different developmental stages spread mosaic-like across the whole lamina.

Variations. Morphological variations of stoma complexes can be expected, since different ways of their development exist. The guard cells of gramineous plants are a typical exception. They are often shaped like dumb-bells.

The hydathodes, that can often be found at the end of vascular bundles are other derivatives of stoma complexes. Their guard cells do still look like those of normal stoma complexes, but they cannot be closed any more. Water secretion by the hydathodes is called guttation. Characteristic hydathodes occur especially at the leaf margins of Garden Nasturtium (*Tropaeolum majus*), Lady's Mantle (*Alchemilla vulgaris*) and at the leaves tips of many grasses. Salt, sugar and organic compounds that are dissolved in the guttation water crysalise after evaporation of the water at the site of outflow. Typical examples are the lime secretions of Saxifrage-species (Saxifraga) and the salt glands of halophytes.

Trichomes

Hairs and scales of plants

Epidermal attachments of various shape, structure and function are called trichomes. They protect and support the leaf, produce glands in the form of scales, different papills and, in roots, often absorbing hairs. They stem exclusively from epidermal cells. Often a trichome is formed by just one cell though sometimes several cells are involved. Trichomes have to be distinguished from multi-cellular trichomes (like spines) and from short shoots (like thorns), since these do contain not only cells of epidermal origin but of other tissues, too.

Hairs can be found at plant surfaces in manifold shapes. They can be single-or multi-celled, branched or unbranched, living or dead. Their walls may be strengthened by silicate, calcium carbonate or other encrustations giving the respective hairs a bristle-like appearance. Such stiff bristles protect plants from being eaten away by animals. Many hairs, particularly the strongly branched ones, help avoiding transpiration losses. It is well known that plants from dry habitats are either succulent or densely covered with silvery hair. 120 hairs per mm^2 were counted at *Stachys lanata* (woundwort). Their microscopic analysis shows strongly branched and dead hair cells. These hairs offer three advantages:

1. The lumen of dead cells is air-filled. It gives them a silvery-whitish appearance. A large proportion of the incoming light is thus reflected. A thick layer of wax has the same effect.
2. Only weak circulation takes place at the leaf surface, thus reducing water loss to a minimum.
3. Dying of the hair cells reduces the surface area, where water could be lost drastically. If the cells were living, too a high loss of water loss would occur due to the largely increased surface of the branched hair cells.

Hairs are organised in a regular pattern on the leaf surface. They are laid out at nearly constant distances. Comparable patterns can be found in the lay out of stomata. Why this regularity? No definite answer can be given until today, but we can present a mathematical model that follows rules, that may very likely be valid for plant tissues, too.

A kaleidoscope of different hairs. Exceptionally long (1-6 cm), single-celled and unbranched hairs out of nearly pure cellulose wrap up the seeds of *Gossypium* (cotton). With *Lobelia, Arabis alpina* or *Malacantha alnifolia,* they are branched. Multi-celled hairs can consist of one or several rows of cells. Typical examples are the dendritic trichomes of the plane tree (*Platanus hybrida*) or those of the dark mullein (*Verbaseum nigrum*). Hairs at the leaves of oaks (*Quercus robur*) look like tufts that of many Malvaceae are stellate. Those of the oleaster (*Elaeagnus angustifolia*) are formed like scales. Bromeliaceae use such scales for water absorption. The humidity of the air is collected within the scales with the help of capillary forces and later on used by the plant. Gland hairs consist of a single-celled stalk and a single- or multi-celled head. Some examples include:

The gland hairs at the leaves of tobacco (Nicotiana tabacum) and e. g. *Adenocaulon* (Compositae) have multi-celled heads, while they are single-celled with primroses (Primula sinensis) and garden geraniums (Pelargonium zonale,for example). The secretion of Pelargonium zonale and *Origanum* (Lamiaceae) is a volatile oil. The gland cells themselves are rich in plasma, secretion takes place through the cell walls. The oil gathers at the cell surface and can be seen as a cap that refracts light strongly and is covered by a thin film of both cuticle and cell wall components. After the cap's splitting-off, the oil is set free. The peel of citrus fruits contains another type of secretory glands that is sunken into the epidermis.

The filaments of Tradescantia virginiana are surrounded by multi-celled hairs. The single cell looks like a barrel. The protoplast lives at the wall and the vacuole is colored with a purple dye from the anthocyane-family. It is crossed by numerous plasma chords. The nucleus is located in the cell's center looking like it were strung up on the plasma chords. Since the plasma chords keep changing their shape all the time, the nucleus is torn to and fro as if at rubber bands. The plasma of the chords has a strong and lively current and transports plainly visible granules. The current is strictly directed and currents of opposing directions can often be seen in other chords. The hairs on the filaments of Tradescantia virginiana are a classic example for currents, since they are easily perceived and very regular in organisation. Consult this site about the molecular mechanisms of plasma currents.

The hairs at the filaments of the hardheads (*Centaurea jacea*) and those of the cornflower (*Centaurea cyanus*) got well-known for other reasons. They are sensitive to touch and cause the filaments to move.

The stinging hairs of the stinging nettle are really multicellular trichomes. They consist of two parts and a multicelled base that develops not only from epidermal cells, but from those of the subepidermal layers, too. A hair cell is sunken into the base. Its basal part is called bulbus. It is surrounded by the cells of the base like liquid is surrounded by a mug. Its upper part is elongated and thin and ends in a laterally attached head. At the site of transition, the cell wall is considerably thinner than in other parts of the cell.

Encrusting silicates make it brittle and cause the head to break off easily when touched. This leaves the hair with a point that bears an astonishing resemblance to the needle of a syringe. The pressure of the touch is directly transferred to the bulbus due to the rigidity of the cell wall. It presses the content of the bulbus (sodium formamide, acetyl choline, histamine) through the cannula and injects it into the wound.

Trichomes Functions of the Epidermis

Papillae are protrusions of the epidermal surface. Typical textbook-examples are the papillae of the pansy flower (*Viola tricolor*) as well as the leaf surfaces of many species from the rain forest. They give a velvet appearance to the plant's surface. Some epidermal cells may be developed into water-storing vesicles.

The water vesicles at the surfaces of many Mesembryanthemum-species and other succulent species are such derivatives. With some plants, like bellflower (*Campanula persicifolia*), the outer walls of the epidermis are thickened like a lens. They collect the light, which is then perceived by specialised light receptors (also called perceptors) and used for physiological reactions.

RHIZODERMIS—ENDODERMIS

The epidermis develops from the outer layer of the apical meristem. The origin of the rhizodermis is much less evident. Depending on the species, it may during evolution either have derived from the root cap or the primary bark.

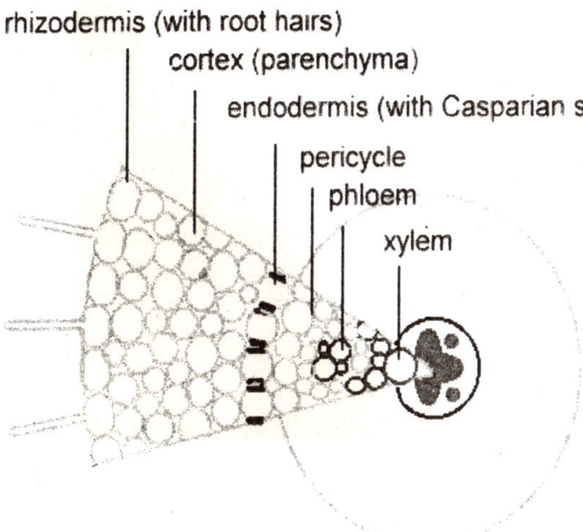

Figure 1. Rhizodermis—Endodermis

The root hairs that absorb water and nutriments are an important feature of the rhizodermis. With some species of the Leguminosae family, they participate in addition in the recognition and uptake of nitrogen-fixing bacteria (belonging to the genera *Rhizobium*). They originate in a zone with high cell division activity. A root hair develops usually from a small extrusion (a papilla) at the apical end of a cell. In many plant species, all cell types of the rhizodermis are able to develop root hairs, while some species have specialised cells (trichoblasts) for this purpose. They are arranged at a regular pattern on the surface of the rhizodermis. The origin and the growth (elongation) of hairs in the different root parts occurs synchronised . The oldest hairs are the longest, while those nearest the root tip are shortest.

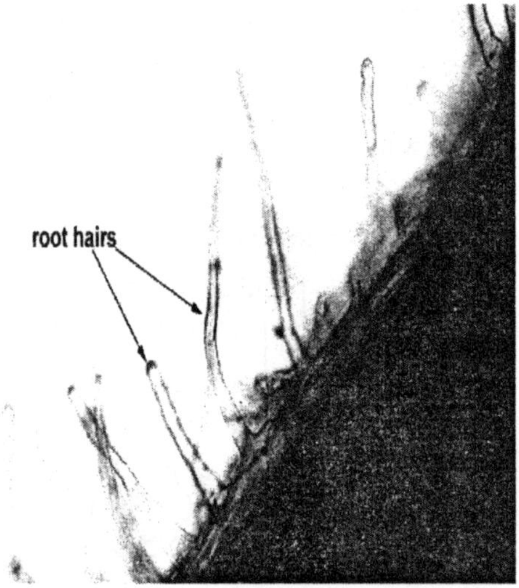

Figure 2. Root hairs of rice.

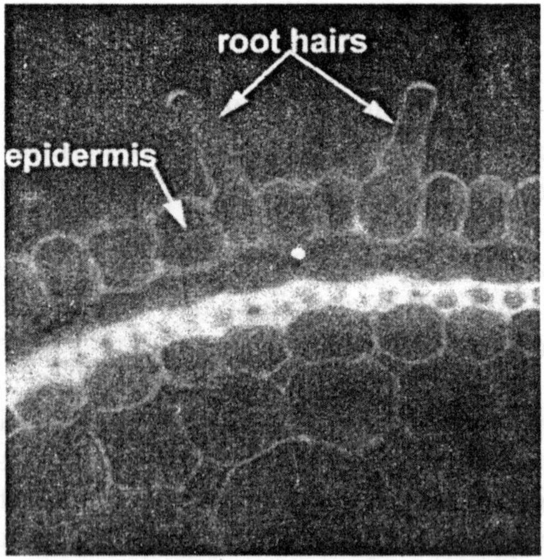

Figure 3. Cross section of an older rice root that is stained with flourestic berberine and with aniline blue counter stain.

Root hairs can become as long as 80-150 μm and as wide as 5-17 μm in girth. They are highly vacuolate when fully differentiated. Only rarely do they consist of many cells (adventive roots of *Kalanchoe*). Root hairs are short-lived (a few days) and are continually replaced by new ones. The formation of root hairs helps to increase the size of the absorbing root surface drastically.

H. J. Dittmer calculated the root surface of a rye plant (*Secale cereale*) in 1937. He counted 13 800 000 roots (including all side roots and branchings) with a total surface of 235 km^2. The roots have 14 milliard living root hairs with a surface totalling 400 km^2. The surface of roots and root hairs thus sums up to 635 km^2. All roots of this one plant were contained in 1/22 m^3. The plant surface above ground covers 1/30th of the root surface.

Even if the cell walls of the mesophyll that border on the intercellular space are taken into account, the absorbing surface of the root is still 22 times as big as the transpiring surfaces of the shoot. This may look uneven, since it enables the plant to a water uptake much larger than its water transpiration. But the calculation includes many elements of uncertainty. It assumes, for example, that all root hairs are equally active. It would therefore certainly be desirable to redo such a calculation with data from another plant of the same or another species to be able to estimate the error rate and to find out whether the calculation can be generalised.

Root hairs are always in close contact with soil particles. They are often grown together and it is thus impossible to isolate a root that has been grown in the soil without impairing it. As a rule, root hairs produce carbohydrate-containing mucigel. It generates a rhizosphere, where nutriments can be processed before uptake by the plant.

Another specialised cell in the root is the endodermal cell. The endodermis forms a cylinder around the vascular cylinder in the roots. The endodermal cells have a suberin layer called the casparian strip embedded into their cell walls. This suberin can be stained using a special dye and appears when viewed under fluorescent light.

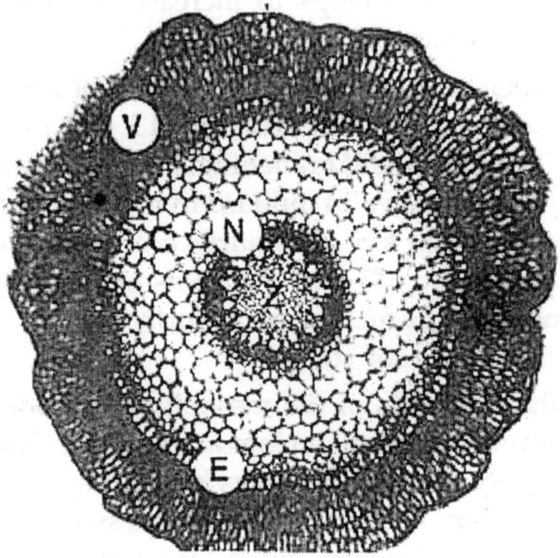

Figure 4. Root and velamen radicum of Dendrobium superbum. *Light microscopic photo of the root.* Velamen *(V), exodermis (E), cortex (c), endodermis (N) and central cylinder (Z).*

The casparian strip appears as light spots in the endodermis in a longitudinal section through the root, but the suberin is laid down in a continuous layer that cannot be seen in longitudinal sections.

The Velamen radicum is a special epidermis occurring only in aerial roots. It is an epidermis of several cell layers that is produced by the aerial roots of Araceae and tree-dwelling orchids. Most of its cells are dead and store water like a sponge. The *Velamen* looks silvery-white when dry, since the cells are filled with air. In a moist state, the chloroplasts of the underlying tissue shimmer through and the roots look green.

SECONDARY DERMAL TISSUES

Secondary dermal tissues are usually found in older shoots and roots, because these are the organs where secondary growth takes place. Only rarely and only in their initial stages do secondary tissues occur in bryophytes, ferns and monocots. They stem from the phellogen, a secondarily developed cambium. The cells of the phellem are called cork cells, they are generated centrifugally, are non-living and have suberised cell walls. The phelloderm consists of cells given off towards the inside of the phellogen, forming the inner part of the periderm. Periderm is the collective term for all three types of tissue.

Cork layers are neither permeable by water nor gas and do only rarely constitute homogeneous structures. The cracked surface structure of the bark that typical for certain species is a visible expression of this. The phellem is often interspersed with groups of parenchyma cells of elongated appearance, called lenticels. These can be observed on the surface of elder (*Sambucus niger*), a typical specimen of basic botanical courses.

The lenticels serve to permit gas-exchange between the metabolically active cells below the bark and the atmosphere. They are not only found in the shoot, but in roots, too. Cork is rarely found within primary tissues, the leaves of *Tabernnaemontana pachysipon* are an exception. They have little islets of cork at the undersurface of their leaves.

The totality of all tissues produced by the phellogen, both primary and secondary ones as well as some other types, is called bark. You could simplify things by saying that all tissue outside the cork cambium is bark.

GROUND TISSUE OR PARENCHYMA

Parenchyma or ground tissue cells are always alive, usually isodiametric and only seldom elongated. The pith of the shoots, the storage tissue of the fruits, the seeds, the roots and other underground organs are all parenchyma tissues, as is the mesophyll (the assimilation tissue of leaves). The mesophyll is reviewed separately due to its fundamental function in assimilation and its resulting importance for plant nutrition.

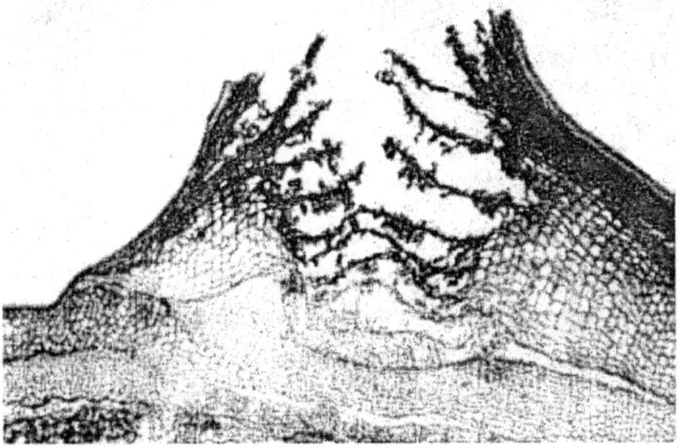

Figure 5. Lenticel from the bark of the tropical leguminosa Entada rhedii *(Mimosaceae)*

Parenchyma cells and relatively large intercellular spaces in a cross-section through the stem of *Geum urbanum*. The preparation has been stained with coriphosphin. The entire function of the parenchyma seems often to be in the filling of the gaps between more specialised cells. This is expressed by the term ground tissue, which means the mass of cells into which more specialised cells, like vascular tissues or seeds are imbedded.

Parenchyma cells are not differentiated, neither morphologically nor physiologically. It is not uncommon that they contain chloroplasts. Sometimes, especially when grown without light, leucoplasts or incompletely structured chloroplasts can be found. Although hardly any cell division takes place in differentiated parenchyma, the cells keep their ability to divide: they make up the bulk of cells participating in wound healing and regeneration of plant organs. Consequently, the cells could be said to be in a kind of waiting position. They form a pool of cells that can be activated when required, for example during ontogenesis and at unusual incidents.

The meristem has already been defined as the tissue, where most cell divisions take place. The parenchyma could analogously be described as a basic tissue, whereof very different tissues develop during ontogenesis. The impact of a cell's position has already been mentioned. Whether or not a cell divides depends on its position (information of position) within the tissue. In the case of parenchyma cells, the position has a decisive influence on their further development.

Parenchyma tissues (from the pith of the shoot, for example), can be cultivated in appropriate synthetic culture media and are thus prompted to divide. They keep their capacity for cell division for several decades, if parts of the culture are transferred to fresh media at regular intervals. The cells stay in an undifferentiated state. They form aggregates, the calli (sing. callus). The cultures themselves are called tissue- or callus-cultures. The calli remain colorless, when kept in darkness, while they become usually green upon light-exposure.. Addition of suitable growth hormones (phytohormones) leads to the development of shoots or roots.

Storage Functions

The metabolism of plants produces a vast variety of substances, some of which are stored for subsequent use. In general, parenchyma cells of different organs above or underground serve as depots. The chemical features of these substances will be reviewed later. It is, at present, sufficient to know that—for cells—basic differences between small and large molecules exist. The majority of small molecules is made up by inorganic ions, organic acids and their salts, sugars, a number of nitrogen-containing compounds (like amino acids or alkaloids) and several others. Small molecules are generally stored as solutes within the vacuole. They are sometimes crystallised and these crystals are either stored within the vacuole or the plasma. Large molecules, or macromolecules, are, for example, starch and proteins. Macromolecules do rarely find a way to the central vacuole. They are stored as microscopic particles within the plasma.

Sugar and starch are the most important primary products of photosynthesis. It is commonly known that the parenchyma of fruits may contain large amounts of different sugars. The simultaneous storage of organic acids in the same cells assigns the fruits their specific, often sour-sweet taste. Starch depots can be found in seeds (within the endosperm, a parenchyma tissue), in roots, tubers (dahlia, potato) and others.

Proteins are frequently stored in seeds, usually in cells specialised for this function. A wheat grain, for example, has starch in its central endosperm cells and proteins in the form of aleuron granules at the periphery. These proteins, the storage-proteins, belong to a certain class of proteins. Their chemical composition and structure is dependent on the respective species.

When germinating, part of the macromolecules of the endosperm are degraded and used for the organisation of the new plant. After mobilisation of its contents, the endosperm usually dies.

Water-Problem

Besides the already mentioned chemical compounds, parenchyma cells store huge amounts of water. This is especially striking in succulent plants, like cactus, *Aloe*, *Agave* and Aizoaceae (Mesembryanthemum family). The water content of single parts of a bamboo shoot and its content of parenchyma cells in percent can be correlated linearly.

Certain organs of plants of our regions do also contain more water than others. Among these are the flesh of fruits, onions, buds and all fleshy parts above ground. The capacity of vacuoles for water-storage is dependent on the molarity of its solutes. It is called osmolarity.

We already got to know the vacuoles of parenchyma cells as depots for a whole range of small molecules. According to the physicochemical conditions just explained, these cells are predestined for the storage of huge amounts of water. This is the reason, why ripe fruits are always firm; the high osmotic pressure may even lead to bursting.

Most parenchyma tissues, particularly those with storage functions have only small, if any intercellular spaces at all. The thin cell walls put up hardly any resistance against the extension of the cell contents due to the high osmotic pressure (turgor) of the single cell, so that intercellular spaces cannot be formed in the first place.

Aerenchymas

The case is different in a special type of parenchymatic cells characterised by thicker walls. By splaying of such cells or tissues, air filled cavities are generated. In the shoot and the roots, these cavities are usually arranged tangentially or radially. The splaying is caused by tension differences in neighboring tissues with different growth rates. When splaying is produced by asymmetrical elongation growth, irregularly formed cells with extensive intercellular systems are created. The spongy parenchyma of foliage leaves is a typical example. Spacious aerenchymas can also be found in the pith of many monocots living in humid habitats like rush. This tissue type is also called star parenchyma because of its characteristic cell shapes.

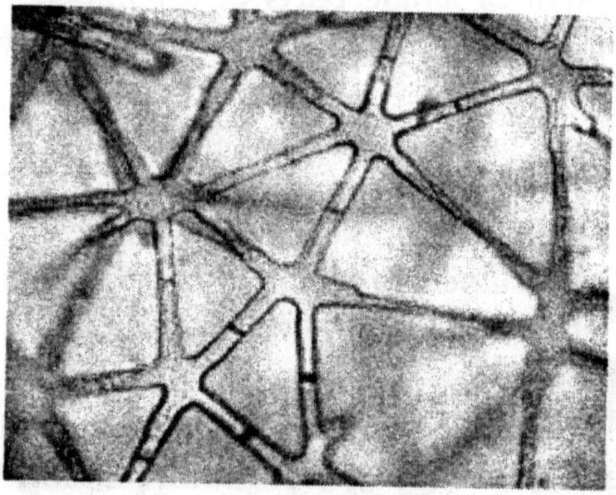

Figure 6. Star parenchyma with extremely large intercellular spaces in the pith of a rush stem (Juncus). *This parenchyma serves as an aerenchyma.*

In the submerged (growing under water) shoots of many angiosperms like the white or yellow water lily exist large air-tubes that can already be seen with the naked eye. They are separated by simple cell layers. This honeycombed construction has two effects: on one hand is a great stability achieved with only little material. This principle has also successfully been applied in technological areas. On the other hand causes the low specific weight a buoyancy that supports the floating of the shoot.

MESOPHYLL

Mesophyll is the main assimilation tissue of leaves. Assimilation tissues are—in a broader sense—all those tissues that are made from chloroplast-containing cells and are thus able to perform photosynthesis. They are found in all the green parts of a plant. An important aspect of photosynthesis is the integration of carbon dioxide into organic compounds. The resulting products are summed up as assimilates. The leaves of higher plants are by far the most important production centers—if you disregard unicellular aquatic algae. Leaves consist usually of three different tissues: the mesophyll, the epidermis and the vascular tissues. The mesophyll is a parenchyma tissue. It is a true assimilation tissue. In the leaves of most ferns and phanerogams, it is organized into palisade parenchyma and spongy parenchyma.

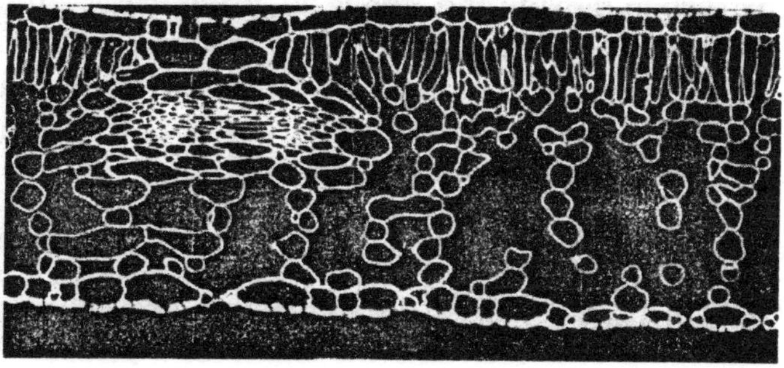

Figure 7. Cross-section through a Helleborus niger *leaf. Semi-thin section of 3 μm thickness*

The "typical" leaf is of a dorsiventral structure. The palisade parenchyma is usually directly beneath the epidermis of the upper surface of the leaf. The spongy parenchyma fills the space beneath the palisade parenchyma. It is interspersed with a voluminous intercellular system, whose cavities are in direct contact with the atmosphere via the stomata. This roughly outlined structure is always cited as being the prime example for selection and adaptation. Our contemporary knowledge of photosynthesis shows that an efficient assimilation tissue has to meet the following criteria:

1. It must make optimal advantage of the available light.
2. The cells have continually to be provided with enough carbon dioxide (CO_2).
3. All cells must be connected via direct cell-to-cell contact with the vascular tissues in order to guarantee water supply and transport of assimilates.

The structure of the assimilation tissue of leaves fulfils these requirements in a nearly ideal way. Additionally, the lamina is thin and flat and the positioning of the leaves at the shoots is designed to take maximal advantage of the light. The shadowing of chloroplasts by other cells or leaves is thus reduced to a minimum. The supply of the plant with light, carbon dioxide and, above all, water varies strongly. It is therefore not astonishing that modifications of this construction with enhanced adaptive values have developed during evolution. Their organisation is dependent on environmental conditions.

Palisade Parenchyma

The cells of the palisade parenchyma are cylindrical. Neighboring cells look like the stakes of a palisade. The analogy is striking, when looking at the cross-section of a leaf, but is less suitable, when thinking of the three-dimensional arrangement of the cells. Cells of the palisade parenchyma contain three to five times as many chloroplasts as those of the spongy parenchyma. The chloroplasts stay usually near the cell's wall, since this adjustment guarantees optimal use of light.

Some species have irregular, or even branched palisade parenchyma cells. Especially ferns, conifers and a few angiosperms harbor such exceptions. No direct relations between these species exist. Does the branched palisade parenchyma have any selective advantage? It is suggested that the resulting enlargement of the surface is advantageous, but then again: why is this structure so rare?

The development and particularly the differentiation of the palisade parenchyma is influenced by extern factors like light and the CO_2-content of the atmosphere. In many species, it is distinguished between sun- und shade leaves. Sun leaves have been exposed to large quantities of light during ontogenesis. This results in a multi-layered palisade parenchyma. Shade leaves in contrast perceive only little light, the palisade parenchyma stays single-layered.

The enlargement of the palisade parenchyma causes usually a reduction of the spongy parenchyma, which is accordingly less well-developed in sun leaves. As important as the exposition to light is the leaf's position at the stem. Old leaves that live near the soil do often have a palisade parenchyma of just one layer's thickness, while younger ones at the top of the plant have normally multi-layered palisade parenchymas. This organisation is most common, but the following variations can also be found:

— palisade parenchyma at the undersurface of the leaf. It is especially striking in scales. Examples: arbor vitae (*Thuja*), leaves of ramsons (*Allium ursinum*).
— palisade parenchyma at both surfaces. This is typical for plants of dry habitats (xerophytes). Example: prickly lettuce (*Lactuca serriola*).
— ring-shaped, closed palisade parenchyma: occurs at cylindrically organised leaves and needles of conifers.

Spongy Parenchyma

The variability of the cells of the spongy parenchyma and that of its organisation is even greater than that of the palisade parenchyma. It is often said to be an aerenchyma, since it is characterised by a large number of connected intercellular spaces. This does not mean that the contact between palisade parenchyma and intercellular spaces is less well developed. On the contrary: the proportion of palisade parenchyma that is in contact with the intercellular space is larger than that of the spongy parenchyma. The elongated and cylindrical shape of the cells enables only selective contact to neighboring cells even at close packing. The spongy parenchyma shows larger contact areas as well as cell-to-cell contacts between palisade parenchyma and spongy parenchyma and between spongy parenchyma and cells of the vascular bundles. This is also the reason for a water and assimilate transport that proceeds without losses.

The intercellular system causes the development of large inner surfaces within the leaf. The ratio of intercellular space to total cell volume is species- and habitat-specific. It ranges around 70-700:1000. F. M. Turrel gained the following results, when measuring and extrapolating the total foliage of a *Catalpa* in 1934:

— inner surface: 5100 m^2

— outer surface: 390 m^2

The ratio of these two surfaces is again dependent on species and habitat. It is low in shade leaves and high in sun leaves.

Evolutionary Trends

Step by step, the two mesophyll types developed during evolution. Both spring from the same ground tissue and numerous transitions prove that the typical palisade and spongy parenchyma cells are merely extremes of a completed differentiation. With few exceptions, all cells of algae are able to perform photosynthesis. The leaves of simple bryophytes are generally single- and only rarely double-layered. No distinction can be made as to epidermis and mesophyll.

Other bryophytes, like the *Polytrichum* -species (mosses) and Riccia-species (liverworts) have lamellar ribs that consist of several layers of tissue on top of each other and are specialised for assimilation. Some liverworts (like Marchantia) have a multi-layered thallus that lives directly at the soil surface. Upon a layer of voluminous parenchyma cells sits an assimilation tissue of numerous small, plump cells gathered together in groups and interspersed with large, air-filled cavities. Liverworts have neither stomata nor guard mechanisms. Instead, their upper surface is mingled with characteristic pores that cannot be closed.

Many aquatic and marsh plants have a homogeneous aerenchym with very large intercellular cavities. They are not structured into palisade parenchyma and spongy parenchyma. Neither can it be distinguished between mesophyll and epidermis. The leaf tissue of the canadian waterweed (*Elodea canadensis*) is reduced to only two cell layers.

The vascular bundles of nearly all vascular plants are surrounded by a bundle sheath, a more or less extensive layer of parenchyma cells. Its cells contain usually less chloroplasts than those of other mesophyll cells. They are frequently also starch-deposits. The cells of the bundle sheath of many -especially tropical- graminaceous plants are enclosed by a wreath of chloroplast-containing mesophyll cells (called Kranz anatomy; Kranz (german) = wreath).

Both cell layers surround the vascular bundles -if seen in cross-section- like two concentric rings. The last years showed that this construction is typical for C_4-plants. C_4-plants are a group of species belonging to different mono- and dicot-families. They integrate carbon dioxide into different organic acids before advancing to photosynthesis. Plants lacking this metabolic pathway are termed C_3-plants. Even within one and the same genus, one species may be characterised as a C_4-plant while another is a C_3-plant. The best known example are the two *Atriplex*-species *A. patula* (C_3) and *A. rosea* (C_4).

There are even some species where the lower leaves are constructed in one way and the upper ones in the other. Accordingly, they have different photosynthesis efficiencies. These species are usually succulent plants living in dry habitats. The stomata are closed during the day to avoid water loss. Hence, CO_2 can only be taken up and accumulated at night (CAM-plants). Fixation is as in C_4-plants; the respective organic acid (normally malate) is stored in the vacuoles of voluminous parenchyma cells lacking chloroplasts. During the course of the day, the carbon dioxide is split off again and, after diffusion into mesophyll cells, fed into photosynthetic pathways.

Development of the Mesophyll

Leaves develop from leaf buttresses, that stem from small lateral bulges of the shoot. The flat structure of the leaf is caused by a preferred plane of division. This type of meristematic activity is attributed to the lamellar meristem, the predecessor of the mesophyll. Differentiation into the typical palisade parenchyma and the spongy parenchyma takes place due to the uneven growth rates of the different cell layers in the developing leaf. The activities of cell division and cell elongation occur in the different layers at different times. Additionally, the cells expand into different directions. An example: the cells of the palisade parenchyma develop by anticlinal division, those of the vascular tissue in parallel. As long as division and elongation of two tissues are concerted, the relation of the cells remains the same. But the cells of the epidermis start rather early during ontogenesis to divide more rapidly than those of the not fully differentiated parenchyma. This causes splaying of the parenchyma cells, whereby the intercellular spaces are generated. They are partially filled again upon further anticlinal division of the palisade parenchyma.

The differentiation of the lower epidermis and the spongy parenchyma is coordinated somewhat more variable. Depending on the species is either the cell division of the epidermis or that of the spongy parenchyma stopped first. Cell division and cell elongation are temporally separated from each other. The different growth rates of the different parts of the developing leaf succeed in a wave-like growth movement. These growth movements were documented in a time-lapse film. It shows impressively how the growth and expansion of a beech leaf spread in intensive wavy movements.

REFERENCES

Barthlott, W.: Epidermal and seed surface characters of plants: systematic applicability and some evolutionary aspects. *Nordic Journal of Botany*, 1981.

Behnke, H.-D., W. Barthlott, "New evidence from the ultrastructural and micromorphological fields in angiosperm classification", *Nordic Journal of Botany*, 1983.

Cormack, R. G. H., "The development of the root hairs by Elodea canadensis", *New Phytologist*, 1937.

Turrell, F. M., "Leaf surface of a twentyone-year-old catalpa tree", *Iowa Acad. Sci. Proc.*, 1934.

4

Vascular Tissues

The architectural design of the plant's body of vegetation is very complex. Thin petioles carry heavy and flat laminas, stems support leaves, flowers and fruits. All plant organs are exposed to mechanical strains. Organs above ground follow the wind's drift. Their high elasticity lets them either return to their original position, or it makes them swing around an imaginative axis. Trunks are stable enough to resist the wind's pulling. They withstand pressure and are inflexible, although their projecting treetops provide the wind with a large target. The wind makes the upper plant organs and the trunk act like a lever, a large part of the force is hence exerted onto the roots, that anchor the plant in the soil. Other functions of the root are water and nutriment uptake.

The strength of tissues protects also against enemies. The hard shell of many seeds prevents a chewing to pieces or puncturing by animals and avoids that parasites like fungi or bacteria force their way into them.

The high water-content of plant cells that lends a high tension to plant tissues and is caused by the turgor. It supplies plant tissues with a certain stability. Its actual importance is seen best in wilting leaves or flowers after their water supply has been stopped. Extensive specialized supporting tissues exist only in vascular plants. Despite the existence of huge marine brown algae (seaweeds, like

Macrocystis, Laminaria), not a single terrestrial alga, whose thallus raises more than a few cell layers above ground, is known. Vascular plants have up to three types of supporting tissue:

1. The collenchyma, a tissue of living cells,
2. the sclerenchyma, a tissue of nearly always dead cells, and
3. the vascular tissue consisting of both living and dead cells. It is responsible for the transport and dispersal of water, nutriments and assimilates.

COLLENCHYMA

The collenchyma is the typical supporting tissue of the primary plant body and growing plant parts. Its prosenchymatous cells are living at maturity and are always kept in a primary state, which means that they are never lignified. Collenchyma walls are interspersed with groups of pits that tend to be organised in special areas. The name collenchyma derives from the Greek word "kolla", meaning "glue", which refers to the thick, glistening appearance of the walls in fresh tissues.

The collenchyma is the typical supporting tissue of the primary plant body and growing plant parts, though it is kept with unaltered structure and function even in outgrown organs like stems, petioles, laminae or roots. In cross-sections of stems, the collenchyma commonly appears as discrete strands or as a peripheral cylinder that lies, depending on the species, either directly beneath the epidermis or is separated from it by several layers of parenchyma.

The cylinder is usually composed of several layers. Collenchyma is also found bordering the veins of dicot leaves. It forms fibres in edgy stems that run along the edges or ribs. Often either phloem or xylem of the vascular bundles is associated with collenchyma cells.

Many transitions prove the collenchyma's origin from the parenchyma. The differentiation is reversible, a degeneration to meristematic states has often been observed. The walls of collenchyma cells are strengthened by the deposit of cellulose and the coating with pectin. These strengthenings are often restricted to single parts or edges of the cell. The walls of parenchyma cells are opened by pits that are often arranged in special areas.

The unevenly thickened cell walls led the German botanist C. Müller to distinguished between different collenchyma types:

1. *Angular Collenchyma.* A thickening of the cell's edges can be seen in cross-section. Longitudinal sections show the elongated shape of both cell and thickening. A cross-section through the stem of *Begonia rex* or related species is the typical specimen used in botanical microscopic courses. Angular collenchyma occurs also in species of the following genera: *Ficus, Vitis, Ampelopsis, Polygonium, Beta, Rumex, Boehmeria, Morus, Cannabis, Pelargonium* and others.
2. *Tangential Collenchyma.* The tangential walls of this collenchyma type are thicker than the radial walls. Examples: *Sambucus nigra,* species of the genera *Sanguisorba, Rhoeo, Eupatoria.*
3. *Lacunar Collenchyma.* While hardly any intercellular spaces exist in the two types above, are those of this type very large. Clear gaps can be recognised between the cells. Occurrence: species of the genera *Lactuca, Salvia, Prunella* and the Composite-family.

The cell walls of collenchyma cells are distortable when stretched. Shape and arrangement of the cells cause a high mechanic stability with a capacity of 10-12 kg/mm^2. This quality is especially advantageous in growing plant organs. It enables the collenchyma cells to stretch in synchrony with the other cells without spoiling the toughness of the tissue.

The new state is stabilised by the simultaneous working-in of additional wall material.

SCLERENCHYMA

The other true supporting tissue is the sclerenchyma. Two groups of sclerenchyma cells exist: fibres and sclereids. Their walls consist of cellulose and/or lignin. Sclerenchyma cells are the principal supporting cells in plant parts that have ceased elongation. Sclerenchyma fibres are of great economical importance, since they constitute the source material for many fabrics (flax, hemp, jute, ramie).

Contrary to the collenchyma, mature sclerenchyma is composed of dead cells with extremely thick cell walls (secondary walls) that make up to 90% of the whole cell volume. The term "sclerenchyma" is derived from the Greek "scleros", meaning "hard". It is their hard, thick walls that make sclerenchyma cells important strengthening and supporting elements in plant parts that have ceased elongation. The difference between fibres and sclereids is not always clear. Transitions do exist, sometimes even within one and the same plant.

Fibres are generally long, slender, so-called prosenchymatous cells, usually occuring in strands or bundles. Such bundles or the totality of a stem's bundles are colloquially called fibres. Their high load-bearing capacity and the ease with which they can be processed has since antiquity made them the source material for a number of things, like ropes, fabrics or mattresses. The fibres of flax *(Linum usitatissimum)* have been known in Europe and Egypt since more than 3000 years, those of hemp *(Canabis sativa)* in China for just as long. These fibres, and those of jute *(Corchorus capsularis)* and ramie (*Boehmeria nivea*, a nettle), are extremely soft and elastic and are especially well suited for the processing to textiles. Their principal cell wall material is cellulose.

Contrasting are hard fibres that are mostly found in monocots. Typical examples are the fibres of many Gramineae, Agaves (sisal: *Agave sisalana*), lilies (*Yucca* or *Phormium tenax*), Musa textilis and others. Their cell walls harbour, besides cellulose, a high proportion of lignin. The load-bearing capacity of *Phormium tenax* is as high as 20-25 kg/mm^2 and is thus the same as that of good steel wire (25 kg/mm^2). But the fibre tears as soon as it is put too great a strain on it, while the wire distorts and tears not before a strain of 80 kg/mm^2. The thickening of a cell wall has been studied in *Linum*. Starting at the centre of the fibre are the thickening layers of the secondary wall deposited one after the other. Growth at both tips of the cell leads to simultaneous elongation. During development do the layers of secondary material seem like tubes, of which the outer one is always longer and older than the next. After completion of growth the missing parts are supplemented, so that the wall is evenly thickened up to the tips of the fibres.

Fibres stem usually from meristematic tissues. Cambium and procambium are their main centers of production. They are often associated with the xylem of the vascular bundles. The fibres of the xylem are always lignified. Reliable evidence for the fibre cells' evolutionary origin of tracheids exists. During evolution the strength of the cell walls was enhanced, the ability to conduct water was lost and the size of the pits reduced. Fibres that do not belong to the xylem are bast (outside the ring of cambium) and such fibres that are arranged in characteristic patterns at different sites of the shoot.

Sclereids are variable in shape. The cells can be isodiametric, prosenchymatic, forked or fantastically branched. They can be grouped into bundles, can form complete tubes located at the periphery or can occur as single cells or small groups of cells within parenchyma tissues. But compared with most fibres sclereids are relatively short. Characteristic examples are the stone cells (called stóne cells because of their hardness) of pears *(Pyrus communis)* and quinces *(Cydonia oblonga)* and those of the shoot of the wax plant *(Hoya carnosa)*. The cell walls fill nearly all the cell's volume. A layering of the walls and the existence of branched pits is clearly visible. Branched pits such as these are called ramiform pits. The shell of many seeds like those of nuts as well as the stones of drupes like cherries or plums are made up from sclereids.

FUNCTIONS OF VASCULAR TISSUES

Vascular tissues do not only fulfil supporting functions. Their conductive functions are more important. They consist of water-conducting xylem and food-conducting phloem. Only the xylem has still supporting functions. Xylem and phloem are combined in a structure that is called vascular bundle.

Vascular tissues transport water, dissolved salts, products of photosynthesis (assimilates), growth regulators like phytohormones and, last but not least, harmful substances and parasites like viruses or mycoplasmas. It is commonly distinguished between the elements of water conduct, the xylem, and those of assimilate conduct, the phloem. Both are combined in vascular bundles. Vascular bundles contain mostly dead cells. In contrast to all other tissues of the actual topic, the phloem, that will be reviewed here together with the xylem has has no supporting function whatsoever.

The lignified cell walls of the xylem elements will at the moment, rather simplifyingly, be called wood. They can last for thousands of years. Their petrified states are ideal evidences for the first occurrence and the successive spreading of the great taxonomical groups of terrestrial plants.

Genuine vascular bundles are the name-giving feature of vascular plants (Trachaeophtes: Pteridophytes and

Phanerogames). Thallophytes have at most first signs of a conductive system, which is not homologous to the respective tissues in vascular plants. Primitive variations of conductive elements appear also in mosses.

The occurring cell types and the arrangement of the vascular bundles within shoots and roots are reliable features for the characterisation of individual classes of plants. The xylem of nearly all angiosperms contains three types of cells:

— wood vessels
— tracheids and
— the already mentioned xylem fibres.

Most gymnosperms lack wood vessels. The few exceptions point out the relationship between gymnosperms and angiosperms. The vascular bundles of monocots seem to run through the shoot without any obvious order, while those of dicots and many gymnosperms are arranged in peripheral rings. Xylem and phloem are separated by a meristematic tissue, the vascular cambium. It provides the means for secondary growth. The vascular bundles of pteridophytes and the roots of flowering plants are located centrally.

XYLEM

The xylem is the principal water-conducting tissue of vascular plants. It consists of tracheary elements, tracheids and wood vessels and of additional xylem fibres. All of them are elongated cells with secondary cell walls that lack protoplasts at maturity. Bordered pits are typical for tracheids, while wood vessels are marked by perforated or completely dissolved final walls.

The xylem takes also part in food storage, support and the conduction of minerals. Xylem and phloem together form a continuous system of vascular tissue extending throughout the plant. The principal conductive cells of the xylem are tracheary elements, of which there are two types, tracheids and wood vessels. Both are elongated cells with secondary cell walls that lack protoplasts at maturity. They are completed by the xylem fibres and parenchyma cells. Much speaks on behalf of the origin of xylem fibres and wood vessels from the tracheids.

Since 1851, the isolation and depiction of lignified cells is done according to the maceration procedure of Schulze. Small pieces of wood are covered with a mixture of potassium perchlorate and concentrated nitric acid. The complete volume should not be larger than 1/10 of the reaction container, because a lot of gas develops very quickly at cautious heating. The surfaces of the wooden pieces are strongly attacked. The single cells can be scraped off after washing of the preparation and examined under the microscope.

There are different cell types. Tracheids are the chief water-conducting elements in gymnosperms and seedless vascular plants. They can also be found in angiosperms. Tracheids are elongated cells, closed at both ends. They are 1 mm on average. Tracheids are regarded as the prototype of prosenchymatic cells, since the cell's ends are pointed and true final walls are missing. Tracheids look often square in cross-section, the lignified secondary wall is relatively thin. Their entire cell surface is evenly coated.

The walls are opened by numerous pits that are, depending on their origin, either round, oval, gap- or groove-shaped. They occur solitarily, statistically scattered, arranged in turns around the middle axis or grouped together. Such groups can often be found at the cell's ends. If gap-like pits lay on top of one another, a ladder- or stair-like perforation may be the result. It is commonly called scalariform. The pits are often surrounded by a halo and are then called bordered pits. Bordered pits are especially common in the tracheids of some gymnosperms. Their structure can be discerned best in a cross-section through neighbouring cells. The middle lamina between the cells is preserved within the pits. Their centre is made up by a disc of primary cell wall material, called torus. No secondary walls exists in the pit's structure. The area between torus and wall (the former middle lamina) is called margo and is very porous, allowing the movement of water and ions from tracheid to tracheid. Bordered pits exist only in cells with secondary walls.

Botanists think of wood vessels (tracheae) as the water-filled tubes of the xylem. M. Malpighi, who thought that he had found an important common element in the anatomy of animals and plants introduced the term trachea in the 17th century. Wood vessels are the chief water-conducting elements of angiosperms.

In contrast to the tracheids the final walls of the single vessels are perforated or, much more so, completely resolved. Wood vessels are therefore generally thought to be more efficient water conductors than tracheids. The length of the single tube makes it difficult to isolate a vessel as a whole. It can be as long as several meters. It is commonly assumed that at least in some species the wood vessels are as long as the whole shoot.

During ontogenesis the wood vessels increase strongly in width.They are usually round in cross-section and have a larger diameter than the tracheids, a feature that enhances their capacity for water-conduct. Exceptionally wide-lumened elements can be seen with deciduous trees, that are known to lose particularly large amounts of water due to transpiration. The water-loss of a fully developed birch tree with an estimated number of 200,000 leaves can be up to 400 litres per day. Even wider are the vessel elements of many lianas.

But the oldest living trees, the redwoods and other sequoias at the pacific coast of California have without exception tracheids with very narrow lumina. Vessels are marked by characteristically structured secondary wall coatings (lignin) at the inner surface of the primary walls. Deposits in the form of screws, rings or nets exist. These strengthenings make it possible for tracheary elements to be stretched or extended, although the cells are frequently destroyed during the overall elongation of the organ.

Beside this vessels with pits or scalariform openings exist, whose walls are nearly completely lined with secondary wall material that is only opened by round or gap-like pits. Many transitions between the two pit types can be found. Often some or even all of the types are members of the same vascular bundle. But there are species, that lack one or the other type. Wood vessels develop -just like tracheids- during primary growth from the cells of the procambium. Where secondary growth occurs wood vessels are produced by the cells of the cambium.

Evolution of Xylem Elements

Tracheids are typical for most pteridophytes and gymnosperms, wood vessels for angiosperms and well-developed gymnosperms, like the Gnetales. Wood vessels of another origin can be found in *Pteridium*, a fern, *Equisetum*, horsetail, and the roots of some *Marsilea*-species. Wood vessels are missing in the primary xylem of some primitive angiosperms. It has already been mentioned that good proof exists for the origin of wood vessels from tracheids. Xylem fibres, too, can be regarded as modified tracheids.

The evolution of multicellular terrestrial plants demanded the development of supporting and vascular tissues. Both functions are best met by elongated cells, but hardly by just one cell type: the supporting function needs the formation of thick and stable cell walls, while the conducting function is best exerted by cells with permeable walls. The first solution to meet this paradox was the juxtaposition of both thickened walls and pits in one and the same cell type (tracheids). This proved to be insufficient, when huge terrestrial plants, like trees, with a high rate of transpiration began to develop.

A new concept evolved. Supporting and conducting functions were separated. Xylem fibres can be regarded as derivatives of tracheids. Their walls are more stable than those of the tracheids, while their conductive function was largely lost; simultaneously the supporting function of the tracheids diminished, while their specialisation into conductive elements was perfected.

Still, the question of the evolutionary relation between tracheids and wood vessels remained to be answered. An alternative to wood vessels stemming from tracheids could be that their origin is unlinked. At the beginning of the 1930th enough proof had accumulated to favour one of the two hypotheses. The Americans F. H. Frost, V. I. Cheadle and I. W. Bailey were able to verify the accuracy of the first assumption. The line of reasoning is based upon logically comprehensible preconditions that have to be met to guarantee a direct genetic link between both structures. At least two of the preconditions can be regarded as fulfilled in the comparative analysis of tracheids and wood vessels:

1. On the assumption that one structure is more primitive than the other and that both are determined by the same genetic program, the derived, more developed structure would have to have properties that are typical of the more primitive structure. If this was not the case, the assumption that both are genetically linked would have to be discarded unless both structures had diverged that much since their separation that all common features were lost. How did the findings and data of tracheids and wood vessels relate to these thoughts?
2. The walls of primitive tracheids like pteridophytes are usually opened by round or elongated pits with either weak or no borders at all. They occur in special areas at the ends of the cells. Similar patterns can be found

in many gymnosperms. Stacked, grove-shaped pits are common. This arrangement is in many ways similar to the scalariform perforation of the vessels. The main difference is the existence of a middle lamina in tracheids that is missing in vessels.

3. Perforation plates are typical for primitive angiosperms, while a complete reduction of the end walls is a feature of further developed taxa. A juxtaposition of both forms is normal.
4. Tracheids reveal no clear end walls. The end wall is also missing in primitive vessels. It becomes ever more distinct during perfection.
5. If a certain feature can be regarded as primitive, then it is very likely that another, closely associated one, is equally primitive.
   - Tracheids are thin, elongated cells, while the cells of wood vessels are short and have a wide lumen. Numerous transitions exist. The longer the wood vessels, the more primitive they are. Their walls have nearly always pits. Pit areas with round pits are looked upon as primitive, scalariform perforations as more advanced and a complete opening as the momentary final stage in the development from tracheids to wood vessels. The widening of the cell lumen is coupled to a strengthening of the secondary cell walls.
   - Bordered pits developed independently from the pits just mentioned. They are constructed rather simple in the tracheids of ferns and angiosperms and are highly developed in contemporary gymnosperms.

PHLOEM

The phloem is the principal food-conducting tissue of vascular plants. Its elements are elongated, just like those of the xylem. In contrast to tracheids and wood vessels, mature phloem elements contain a protoplast and sometimes even a nucleus. Phloem elements may be of primary or secondary origin though the early primary phloem, the protophloem, is frequently destroyed during elongation of the resepctive organ.

The main conducting elements of the phloem are the sieve elements, of which there are two different types: sieve cells and sieve-tube members. Sieve cells have narrow pores, their sieve areas are quite uniform in structure, and they are distributed evenly on all walls. One of the principal differences between sieve cells and sieve-tube members is the presence of sieve plates in sieve-tube members, that are absent in sieve cells. Sieve cells are the only type of food-conducting cells in most seedless vascular plants and gymnosperms, whereas in angiosperms only sieve-tube members are present. Sieve-tube members occur end-on-end in longitudinal series called sieve tubes. They are in contact via plasmodesmata. Typically, the final walls are interspersed with primary pit areas (groups of plasmodesmata), that later on develop into sieve plates. Sieve tubes in the phloem of angiosperms are flanked by one or several plasma-rich, nucleated companion cells, that do not occur in gymnosperms.

Much less is known about the phylogeny of sieve elements than about that of xylem members. One cause are the cell walls of sieve elements, that are strengthened entirely by cellulose and are therefore not as resistant as lignified ones, another is the transitory working order of the single sieve element. After loss of function the cells are reorganised and lose their typical structural properties. Good fossils are thus rare.

Despite these obstacles it was possible to reveal in comparative plant anatomical studies (of monocots) definite phylogenetic trends:

1. The originally scattered pits are concentrated in 'sieve areas' (accumulations of pores). The diameter of a pore is around 0.1 - 15 µm.
2. Specialised sieve areas are concentrated in the final walls.
3. A gradual change of orientation of these final walls from very sloping to right angles occurs.
4. The change is followed by a stepwise transition from compound to simple sieve plates.
5. The activity of the sieve areas in the side walls is reduced. This leads to canalising of the flow of assimilates in longitudinal direction and the long-distance transport rates are enhanced.

Phloem elements of the roots are normally organised in a more progressive and thus more efficient way than their equivalents in the shoots. Callose is deposited in the sieve plates at regular intervals and can be detected with special reagents. Resorcin blue gives a blue, aniline blue a bright yellow staining. The amount of callose increases with cell age, continuously reducing the diameter of the pores. Sieve elements that have lost their function are blocked by thick plugs of callose. The question, what is cause and what result remains. Are the cells inactivated by the callose plugs or do these form as a result of the loss of function?

Active sieve tubes contain huge amounts of so-called P-proteins (phloem - proteins). It has often been speculated, whether they have an active part in assimilate transport. No answer has been found up until now.

The sieve tube elements of angiosperms (mono- and dicots), but not those of pteridophytes and gymnosperms are associated with companion cells. In *Ginkgo* and other gymnosperms their function is taken over by specialised parenchyma cells, that resemble companion cells in structure: the albuminous cells. They work out the contacts between phloem and the surrounding transfusion parenchyma and can also be found as mediators between ray parenchyma cells of the bark and mature sieve cells. Albuminous cells participate in loading and unloading of these cells. The process has been studied in detail in *Pinus*.

Companion cells and sieve elements are of meristematic origin, their development is similar to that of the xylem. Primary phloem develops by longitudinal division and subsequent elongation of meristematic cells. The cells do often divide unequally. The bigger daughter cell differentiates into a sieve element, the smaller after one or two further longitudinal divisions into two to four companion cells. Consequently several companion cells may belong to one sieve cell. The exact number is usually typical for the respective tissue, but it may even vary within a single plant.

Beside the typical phloem elements, fibres and/or sclereids as well as parenchyma cells that serve as deposits for starch, fat, oil and other nutrients, can belong to the phloem. The deposits contain often tannic acids and resins.

VASCULAR BUNDLES OF MONOCOTYLEDONS

Different types of vascular bundles evolved in the different plant groups. Monocots have usually collateral vascular bundles, a type that is also called closed. The vascular bundle is enclosed by a bundle sheath of parenchyma cells.

Xylem and phloem of the shoots of mono- and dicots are usually arranged opposite to each other in the vascular bundles (collateral vascular bundles). The xylem is normally at the inside, the phloem at the outside. In dicots, both components are separated by the fascicular cambium. It is also spoken of an open vascular bundle in contrast to the 'closed' bundles of monocots that lack the cambium.

Xylem and phloem are surrounded by a bundle sheath of parenchyma that is often starch-containing. The prime object for the demonstration of structure and arrangement of monocot bundles is the shoot of corn (*Zea mays*). Although the vascular bundles seem to be scattered in cross-section, do they occur at the periphery in larger numbers. Central bundles are usually bigger than peripheral ones.

Many monocots have hollow stems. It seems as if their vascular bundles were arranged in a ring, because space is rare. In the stems of some aquatic monocots *(Elodea, Potamogeton)*, they are united in a central (axial) strand. This arrangement is of secondary nature, since it developed as an adjustment to aquatic life. A comparable structure can be found in some submerse dicots. It decreases the strength of the stem, while at the same time increasing its flexibility. Exactly this property is needed to meet the demands of running water. Selection favouring efficient water transport cannot be expected in submerse living plants anyway.

VASCULAR BUNDLES OF DICOTYLEDONS

The collateral open vascular bundles of dicots display xylem and phloem that are separated from each other by the fascicular cambium. These vascular bundles are also often surrounded by a bundle sheath and the whole structure is imbedded in a parenchyma tissue called pith. Another type of vascular bundles are the bicollateral bundles. Vascular bundle and interfascicular cambium (*Ajuga reptans*)

For reasons of simplicity, the organisation of primary vascular bundles in the shoots of herbaceous plants is discussed first. The vascular bundles of herbaceous plants are termed collateral and open, because xylem and phloem are opposed and separated by a fascicular cambium of usually several layers thickness. There are exceptions from this organisation: collateral open vascular bundles exist also in dicots, some monocots have collateral closed and some gymnosperms collateral open bundles.

The circular arrangement of vascular bundles can be observed very well in cross-sections of the stem. Interfascicular cambium occurs in many, but not all species (for example the mentioned *Ranunculus repens*). Herbaceous dicots do always have a fascicular cambium, that is, however, usually not divided any more.

The structural elements of xylem and phloem are the same as in monocots. The vascular bundles are often enclosed by a bundle sheath of cells with thickened walls and are embedded in a parenchyma of large cells, the pith. Parenchyma areas between the bundles are often called ray parenchyma. Deviations of this organisation are found in square stems, where the arrangement of the vascular bundles and the shape of the cross-section are in tune with each other.

A largely changed pattern appears in some families (Solanaceae, Asclepidaceae, Compositae, Curcubitaceae and others), whose vascular bundles are open and bicollateral, i.e. the xylem is on both sides associated with phloem. The vascular bundles are arranged in two concentric circles, the lumina of the inner circle being much wider than those of the outer.

Aristolochia sipho has proven useful for the demonstration of secondary growth and thus also for the development of wood. Very young branches have merely any fascicular cambium. Interfascicular cambium develops very early during the first year even before the beginning of secondary growth. This pattern is largely kept in branches of several years' age. The activity of the cambium increases their diameter and the vascular bundles become elongated in cross-section. In general, far more xylem than phloem elements are produced. Annual rings are clearly visible, because at the beginning of each vegetation period (in spring), vessels (conducting function) and fibres (supporting function) with a wide lumen are assembled first, the so-called early wood. In the following, elements with steadily narrowing volumes are produced. In autumn, only few vascular elements with narrow lumina (late wood) form.

GYMNOSPERM WOOD

Gymnosperm wood is more homogeneous than that of angiosperms. Its only conducting elements are tracheids. Just like the wood of angiosperms, it is interspersed with radially orientated rays (medullary rays) consisting of parenchyma cells and sometimes also tracheids.

The wood of gymnosperms is simpler and more homogeneous than that of angiosperms. Except for the species of the order Gnetales, tracheids are the only conducting elements in gymnosperms. The most important insights into the structure and formation of pine wood that is always regarded as the prototype of gymnosperm wood stem from the German Karl Sanio (1832-1891). His observations were completed by the studies of the American botanist I. W. Baily (1954). Sanio assumed (1872/73) that "... the bast (phloem) and wood cells of one radial row develop from a single cell of the cambium by alternate divisions". This observation is one of the first and most important concerning the cambium's impact on secondary growth. The processes within angiosperm wood are in principal the same, but it was at first by no means simple to recognise the logical connection. *Pinus silvestris* or related species like the North American white pine (*Pinus strobus*) turned out to be ideal specimens. Pine wood is a typical object of every botanical ground course. Since cells and tissues are three-dimensional objects, they do have to be regarded from three sides to get an impression of their spatial organisation:

— cross-section

— radial longitudinal section

— tangential section.

These three perspectives can be combined in a diagram. For some years now, scanning electron microscopy has been used in the examination of the three-dimensional structure of wood.

The maybe best-known property of gymnosperm- and angiosperm wood beside the annual rings is the grain, that is visible especially nicely in tangential sections. The veneers often used in the fabrication of furniture are made from such sections.

Annual rings> are well-suited for the determination of the trees' ages. Their thickness depends on many factors. During the 1980th, the ^{14}C method has been established. It is based on the fact that in every carbon-containing compound (like lignin) not only the normal carbon isotope ^{12}C is found, but also, in much lower amounts, the radioactive isotope ^{14}C. The $^{14}C/^{12}C$ ratio of the atmosphere is $1{:}10^6$. ^{14}C has a half-life of 5770 years. No new carbon is incorporated into an already finished compound so that its ^{14}C content sinks continuously. The ratio of $^{14}C/^{12}C$ shifts thus in favour of ^{12}C. Some several thousand year old Californian *Sequoia*- and *Pinus aristata*-trees turned out to be the ideal specimen to test the precision of both methods and to calibrate one with the help of the other.

The xylem of gymnosperms has few or no parenchyma cells at all. Their existence (or non-existence) is a feature of certain genera. With *Pinus,* they are found only in the epithelia of resin ducts, with many Podocarpaceae, Taxodiaceae and Cupressaceae (cypresses), they are amply present, while they are missing completely with Araucariaceae and Taxaceae (yews).

The tracheids of gymnosperms are 0.5-11 mm long and are orientated along the shoot or root axis. They border at neighbouring tracheids above and below not with their final walls, but with the ends of their lateral walls. This is the reason, why there are never, neither in tracheids nor in vessels, ideal, vertical conducting tubes.

In many species, it is distinguished between sapwood and heartwood. Sapwood is an active, water-conducting tissue, while heartwood is inactive and has only supporting functions. Its cells contain but little water or reserve compounds. Instead, organic compounds, oils, rubber, resins, tannic acids, dyes or aromatic compounds are stored here. Oxidised phenolic compounds give the wood a dark colour. The typical heartwood is missing in spruce (*Picea excelsa*), pine (*Abies alba*) and in some angiosperms (poplar, willow). These woods are regarded as less valuable by the timber industry than heartwood-containing ones.

Medullary rays Wood is at regular distances interspersed with radially orientated parenchyma cells and often also with tracheids. They spring from the ray initials> of the cambium. The medullary rays of conifer wood are normally only one cell layer thick, but they can be as high as 1-20 (sometimes even up to 50) layers of cells. Ray tracheids and axially orientated tracheids are connected *via* pits. Medullary rays with resin ducts seem spindly in tangential sections. Resin ducts can be either axially or radially orientated. There is actually hardly any difference between them and intercellular spaces that have been enlarged by the splaying of parenchyma cells. Consequently, resin ducts are always lined with parenchyma cells. Their existence has different reasons: injuries by frost or winter damages are some factors that stimulate their formation. The respective gymnosperm families react differently to these disturbances.

ANGIOSPERM WOOD

The structure of angiosperm wood is more inhomogenous and variable than that of gymnosperms. It can be of two different vessel arrangement patterns: it belongs either to wood with statistically scattered pores or to wood with ring-shaped pores. Angiosperm wood does also contain rays. They are often considerably larger than those of gymnosperms. Monocotyledon wood is characterised by a building plan that differs completely from that of the gymnosperms or dicots. This difference results in a strongly enhanced flexibility.

Angiosperm wood in its typical structure is found almost only in dicots. First signs of wood formation occur in some monocot species, but their wood is always built inhomogenously. Contrary to the very clear architecture of gymnosperm wood, that of angiosperms is more complex and variable, partly because of a greater number of different cell types in the axial system. Most wood contains vessel members as conducting elements and, they can be structured very differently. Over and above the wood may hold a high proportion of xylem fibres and parenchyma cells (the latter constitutes up to 23 percent of all xylem elements). The organisation of rays varies also, they include almost only parenchyma cells, but parenchyma of strongly varied shaping. The differences in shape and orientation of xylem elements can be explained by the organisation of the cambium.

No plant species with 'typical' angiosperm wood exists, rather can the following variables be consulted as classification and identification features for different woods:

— existence or non-existence of vessels
— distribution of the vessels within the tissue
— shape, sise and arrangement of rays
— distribution of axial parenchyma
— construction type of the vessels' perforation plates

Environmental influences lead to manifold modifications, the composition of the wood changes with increasing age. Two main patterns of vessel arrangement have been recognised:

1. Wood with statistically scattered pores: All vessels have about the same diameter, they are evenly

distributed in every new annual ring. Examples are maple (*Acer*), birch (*Betula*) or *Liriodendron*, a Magnoliaceae.

2. Wood with ring-shaped pores: The vessels have diameters of different size, such with large ones are normally laid out in spring. In autumn, almost only xylem fibres are laid out. This pattern is considered highly specialised. It has been shown to exist in a few species of the northern temperate zone. Chestnut (*Castanea*), ash-tree (*Fraxinus*), *Robinia* and certain oak species (*Quercus*) are examples.

Besides the clearly developed types, a number of transitional structures exist. Wood with ring-shaped lumina conducts water nearly only in the outer growth zone. In spring, the lay out of vessels is speeded up and the flow of water is ten times as high as in wood with scattered pores.

Many tropical species develop no annual rings. 75 percent of the trees of the Indian rain forest, 45 percent of those of the Amazonas Basin and 15 percent of the Malaysian trees have no annual rings. Plants of dry habitats, that are in permanent contact with the ground water like *Tamarix articulata, Acacia raddiana* or *Acacia tortilis* show steady activity throughout the year. Transplantation experiments revealed that the rhythm of cambium activity is inherited in some species, while it is chiefly controlled by extern factors in others.

The vessels are connected laterally (exchange of solutes *via* pits). In some species, the contact is restricted to vessels in growth zones, while it extends over several vessels in others. Ample xylem fibres are common in most angiosperm woods. In wood with ring-shaped pores, they are predominant in late wood. In some species, the xylem fibres contain living protoplasts that serve generally as starch deposits. This type of organisation is estimated progressive. Lacking or small numbers of axial parenchyma cells are strikingly correlated with this type of organisation.

Rays of angiosperm wood contain only rarely tracheids, but are constructed from parenchyma cells, whose longitudinal axis is either orientated radially or axially. Both single layered and multi-layered rays occur. The rays of dicots are often considerably larger than those of gymnosperm wood. They range from one to many cells wide and from one to several hundred cells high.

Just like the wood of gymnosperms, that of angiosperms is interspersed with numerous intercellular spaces, whose structure, development and content are, however, more variable. Besides channels, cavities containing resins, oils, rubber or mucigels occur often. Their walls may or may not be lined with parenchyma cells. The intercellular spaces are, just like in gymnosperms, formed by splaying or ripping. Rubber-like substances are often secreted only after degeneration of the cells. This may also lead to filling and subsequent loss of function of neighbouring cells. The production of rubber can be caused by infections, damage due to insects or physiological disorders.

Structure and arrangement of xylem elements determines the strength and other physical properties of the wood and thus its economic value. An important, though not exclusive, indication of the strength is the specific weight. The specific weight of dry, water-free wood is dependent on the mass of the cell wall material per unit volume. The weight of pure wall substance is 1.4-1.62 g/cm^3; the same substance measured within the intact tissue has a weight of 0.04-1.4 g/cm^3. The value 0.04 was measured for the wood of a leguminosae (*Aeschynomene*), while the value 1.4 has been found in a species of the Rhamnaceae family (*Krugiodendron*). The difference is founded on the way the tissues are constructed and on the relative amounts of the individual xylem elements.

Xylem fibres with thick walls and small lumina add to the increase of the specific weight, wide lumina and thin walls decrease it. On the other hand, the strength is amplified by xylem fibres and weakened by vessel elements. Wood with ring-shaped lumina is less resistant to some strains due to the concentration of the vessels in defined areas than wood with diffusely dispersed vessels. The famous and astonishingly stable balsa wood (*Ochroma pyramidale*, a species of the Bombacaceae family) has a specific weight of 0.1-0.16 g/cm^2. This low weight is partly caused by the high proportion of large, thin-walled parenchyma cells.

Two general principles of construction underlie light wood. Either layers of thick- and thinwalled cells alternate or both cell types are evenly distributed and form a structurally homogeneous tissue.

Monocotyledon Wood

The structure of monocot wood is fundamentally different from that of dicots. Vascular bundles strengthened by extremely tough fibres with small lumina are embedded into voluminous parenchyma that consists of regularly arranged cells with large lumina. The wood of monocots is built on a 'reinforced concrete principle' that lends it a great strength of flexibility. Just think of bamboo or the high, slender palms that resist even the strongest tropical storms, or the liana-like rattan cane, whose wood is used for the production of rattan furniture. Gymnosperm wood, however, is constructed on the 'brick principle'. It is thus not astonishing that the wood of spruce or pine splinters easily.

Vascular Bundles of Roots

The vascular bundles of roots are enclosed by two concentric tubes of cells, the pericycle and the endodermis. The whole complex is embedded in the primary bark.

Casparian strips are a typical feature of the vascular bundles of roots. Roots are able to perform secondary growth, which starts with the development of secondary cambium and leads to a restructuring of the vascular bundles. A feature of roots with secondary growth are feeder roots.

The vascular bundles of roots are constructed rather simply and show but little variations, if compared to those of the shoot. One reason could be the less variable surrounding that poses fewer demands than the surrounding above ground. Xylem- and phloem elements are combined in a central strand without pith, the stele. The root of dicots is often regarded as the basic type. The root of monocots differs only negligibly as far as the primary construction is concerned so that a separate reviewing is unnecessary.

The roots of many dicot species like wood plants or perennial plants are able to perform secondary growth, while roots of monocots can, with few exceptions like *Dracaena*, not grown in this way. The primary cylinder of vascular bundles is as a rule composed of radially located xylem elements and peripheral phloem. The vascular elements are enclosed by two concentric tubes of parenchyma tissue, the pericycle (inner cylinder) and the endodermis (outer cylinder). The pericycle is missing in the roots of some simple aquatic plants and in some parasitic species. The whole complex is embedded into a parenchyma of large cells the primary bark.

The radial walls of the endodermis are marked by characteristic thickenings, the Casparian strips that were discovered by Robert Caspari, a German botanist, in 1865/66. These thickenings contain lignin, suberin and other encrusting substances, among them sometimes also phenolic oxidation products that lend the strips a dark colour. The endodermis is regarded as a typical cell layer of roots, but it is frequently found in shoots, too, where it, just as in roots, is localised at the periphery of the vascular cylinder. In the shoots of some ferns surrounds the endodermis single vascular bundles.

The relation of the vascular bundle arrangements of roots and shoots is illustrated by the theory of steles. It is based upon the assumption that root and shoot are merely two parts of one axial unity. A continuously recurring problem is the question of the vascular bundles' reorganisation in the zone of transition between both organs. It should, strictly spoken, not be talked of a root to shoot transition, since the two organs are separated by both epicotyl and hypocotyl. These are formed first in the development of the embryo and only later are shoot and roots laid out. During this early phase, vascular bundles are formed in the area of the epi- or hypocotyl that elongate, beginning at the point of initiation, simultaneously both in the direction of the shoot and the roots and in this process adopt the organ-specific arrangement.

Secondary growth occurs, as has already been mentioned above, in the roots of many dicots. In the course of this, as a rule a secondary cambium develops out of parenchyma cells that have been laid out during primary growth between the xylem rays. The cambium is supplemented by cells of the pericycle so that finally a homogeneous cambium cylinder is formed. As a result of the secondary growth, pericycle and epidermis tear apart. The consequent gaps are filled by division of additional parenchyma cells. Just like shoots, roots develop a secondary final tissue (phelloderm, phellogen, phellem).

Although the vascular bundles of roots and shoots share many principles of organisation, some fundamental differences exist. The formation of the shoot's vascular bundles is normally linked to the simultaneous lay out of leaves or branches and starts at the apical meristem. Lateral roots spring from cells of the pericycle and grow from the inside towards the outside. Only in a second, independent step of differentiation is the contact of the newly developed vascular bundles to the primary (tap) root established. The lateral roots may branch again. In this way, side roots of 2., 3.,.... order develop.

Contrary to leaves, side roots are built with central symmetry in a strictly morphological sense (exceptions are, for example, the roots of *Kalanchoe*), while leaves and leaf stalks are normally organised dorsiventrally. But a physiological dorsiventrality seems also to exist in lateral roots.

It is known that roots perform two in some respects contradictory tasks. On one hand they anchor the root in the soil, on the other hand they take up water and nutrients from the soil. The anchor function demands a tension-proof root, interspersed with strengthening elements like lignification or fibres within the phloem. This function is supported by secondary growth. On the other hand lignification and formation of bark decreases the ability for water uptake and transport. Therefore has, also in roots, the 'division-of-labour-principle' been put through. Perennial plants in particular have at the endings of strongly lignified lateral or primary roots unlignified, short-lived branchings, called feeder roots. Additionally the secondary final tissues are not completely water-tight. How far they are engaged in the uptake of nutrients is not yet known.

Beside the primary and lateral roots, many mono- and dicots are equipped with so-called adventitious roots. They originate in different parts of the shoot or root. The root system of many Poaceae, for example, consists nearly entirely of adventitious roots. Often are roots or parts of them including the hypocotyl used as storage organs. The parenchyma tissue of the primary bark, the xylem and the phloem swell up and numerous cell divisions take place. The newly formed deposits are secondarily supplied with vascular bundles.

SPATIAL ARRANGEMENT OF VASCULAR BUNDLES AND STELES

The spatial arrangements of vascular bundles is explained by the theory of steles. The term stele sums up the vascular system, associated tissues and the enclosed pith. Different types of steles, like the protostele, the plectostele or the siphonostele exist. Monocot steles are a highly specialised exception.

The hitherto expounded explanations about the structure of vascular bundles are based upon the evaluations of cross-sections of roots and shoots. These preparations are suitable for the depiction of the position and the organisation of the vascular bundles (location of the xylem, phloem, cambium, etc.), but they give no hints as to the spatial arrangement and the development of the vascular bundle system. It is rather difficult to get a clear picture here. Serial sections and the derived reconstructions of the vascular bundles' shapes provide a necessary, but very time-consuming method. Another possibility is given by the maceration of the stem, a controlled rotting leading to the assumption that the vascular bundles are the most robust tissues and will therefore survive the maceration procedure rather unimpeded. Despite methodological adversities, the mentioned methods have been used successfully. Today, we know the vascular bundles' spatial arrangements of manifold plants from the most diverse classes; additional information has been gained by the evaluation of fossils.

The term stele sums up the vascular system, associated tissues and the pith. It stems, like so many other terms of plant anatomy, from the Greek and means pillar. The theory of steles says that the primary plant body of roots and shoots is constructed according to the same principle, since both contain a central pillar (stele) that is embedded into primary bark. The theory of steles does not deal with secondary growth. The pattern of steles can be regarded as a conservative feature, whose complexity increased in the course of evolution.

The protostele is a conductive system of simple organisation. It is a simple, unbranched, centrally located axial strand of xylem coated or interspersed with phloem. A protostele contains no pith. It can be found in primitive vascular plants, like the fossil *Rhynia,* as well as in young fern shoots, in the shoots of simple aquatic plants and in the roots of most angiosperms. A variation, the plectostele is typical for Lycopodiae.

The siphonostele represents the prototype of the vascular systems of ferns and all further developed vascular (seed) plants. It is composed of several axial vascular bundles that are arranged within the stem in the shape of a tube with enclosed pith.

Leaf and branch traces branch from the axial bundles in direction of leaves and branches and extend into these organs. Leaf traces start always at the point of branching and stretch out to the basis of the petiole. A leaf can be supplied by one or more leaf traces. The axial bundles run along the whole length of a stem section. If you recall how vascular bundles develop during ontogenesis you will understand immediately that they consist of segments that were originally laid out as leaf traces.

The transition from leaf trace to vascular bundle happens as soon as an apically laid out leaf trace gets in contact with the existing vascular system. Axial bundles are usually interconnected thus creating a communicating conductive system. The areas, where the leaf traces branch, are the nodes, the in-between sections are called internodes. The structure of the nodes in dicot stems is rather simple, complex linking occurs chiefly in monocots.

Some examples of the organisation of vascular bundles will follow. Simple conditions are found in primitive, fossil gymnosperm predecessors (progymnosperms), like in the species *Callixylon brownii*, whose vascular system was reconstructed by C. B. Beck in 1979. It reveals mostly parallel bundles with only few linkings (anastomoses). A considerable increase in the complexity of the branching pattern occurs in gymnosperms. It is distinguished between open and closed vascular systems, the latter are the leaf traces interconnected by anastomoses. Each leaf of *Ginkgo biloba* is supplied by two leaf traces that stem from different, but neighbouring axial bundles. The Cycadales like *Dioon spinulosum* have five leaf traces per leaf that spring from distant axial strands. The leaf traces enclose the cylinder of vascular bundles like a ring, before they branch together in the direction of the leaf. Here, the organisation of the nodes is already rather complex.

Angiosperms stand out due to the high variability in the construction of their vascular cylinders, although the basic patterns are mostly identical. Nevertheless a serious difference between mono- and dicots does exist. The situation in dicots is closer to the basic pattern and is therefore discussed first.

Variations of the vascular cylinder are based upon direction and height of the helix, upon the number of traces per leaf and the nature of leaf insertion. Angiosperms do as a rule contain five strands of vascular bundles. This number is regarded as a basic unit, of which variations with more or less chords are derived. 91 percent of all species have open systems, closed ones, where several leaf traces and leaf trace complexes are connected are typical for the others (both herbs and wood plants). Here, the basic pattern shows four vascular bundles per vascular chord. Often the leaf gaps are very big, like those of many angiosperms (cacti, Cycadeae, ferns). Their vascular cylinder looks net-like.

The following trends can be determined in the evolution of seed plants:

1. Open vascular systems are primitive, closed ones derived.
2. The unilacunar type (just one leaf gap) is primitive, multi-lacunar ones are regarded as derived. Exceptions, transitions and degenerations can be found in many dicots. Often two or more types can be found within one plant. The unilacunar node system is the most common type among primitive seed plants.
3. Vascular systems with five traces are most likely primitive, those with more or less traces are derived.
4. Steles with leaf traces that run through one or a few internodes before branching into the leaf petioles can be regarded as primitive, those that run through many internodes as derived.
5. A helical arrangement of the leafs of seed plants is regarded as primitive.

Monocots provide a highly specialised exception. The vascular bundles of monocots seem to be scattered across the whole stem, being slightly more concentrated at the periphery. If the course of a single axial bundle is examined, a clear helical arrangement can be detected. Never is a vascular bundle exclusively central or peripheral. Depending on where the stem is cut, it can sometimes be found in the centre and sometimes in the periphery. At regular distances leaf traces branch. They, too, run for a long time along the stem, though normally in the periphery. This is also the reason, why particularly many vascular bundles can be seen in the periphery of cross-sections. Part of them are axial chords, part leaf traces. They can usually not be distinguished in cross-section, although axial chords have a slightly greater diameter. Axial bundles may be interconnected by bridging bundles.

REFERENCES

Beck, C. B., R. Schmid, G. W. Rothwell, "Stelar morphology and the primary vascular system of seed plants", *Botany Revolution*, 1982.

Carlquist, S., *Comparative plant anatomy*, New York: Holt, Rinehart and Winston, 1961

Cheadle, V. I., and N. W. UHL, "Types of vasoular bundles in the Monocotyledonae and their relation to the late metaxylem conducting elements", American Journal of Botany, 1984.

Cheadle, V. T., "Research on xylem and phloem—Progress in fifty years", *American Journal of Botany*, 1956.

Esau, K., "Origin and development of the primary vascular tissues in seed plants", 1943.

Foster, A. S., "Comparative morphology of vascular plants", San Francisco: W. H. Freemen and Comp., 1959

Skipworth, J. P., "The primary vascular system and phyllotaxis in Hectorella caespitosa", *New Zealand Journal*, 1962.

Zimmermann, M. H., P. B. Tomlinson, "The vascular system of monocotyledoneous stems", Botany Gazatte, 1972.

5

Intracellular Structures

Almost all intracellular particles have a size at or below Abbe's resolution limit. Though numerous structures like the grana of chloroplasts or the mitochondria were already known or recognised before the use of the electron microscope, no closer scrutiny was possible and only few new insights were made.

Since electron microscopic preparations are very thin compared to the thickness of the whole cell, only one plane of the cell can be seen at a time so that it becomes necessary to check a number of slices to gain a realistic impression of the three-dimensional organisation of intracellular structures. It was therefore not before extensive series of images had been prepared that models of the actual three-dimensional structures of the intracellular contents could be developed.

Many structures, that can be seen through an electron microscope are images of macromolecules or macromolecular complexes. All of them can nowadays be analysed with methods of biochemistry or molecular biology and we know by now much more about their molecular structure and function. Their function will be talked about elsewhere .

These days, the electron microscope is often used as an aid for the examination of the homogeneity of certain cell fractions like, for example, mitochondria, isolated DNA or genes, etc. It is impossible to analyse living cells with the electron microscope, because the preparations have to be contrasted and afterwards to be examined in a vacuum.

All eucaryotic cells are interspersed with extensive membranous systems that structure the cell's interior into numerous compartments. The membranes are of uniform thickness and consist of a lipid bilayer. Among the membranous systems of the cell are the plasmalemma or plasmamembrane, the Golgi apparatus and the endoplasmatic reticulum (ER). The ER forms a continuous system with the nuclear envelope. The latter is composed of two lipid bilayers perforated by the nuclear pores.

Organelles are membrane-bordered functional unities of the cell. Both mitochondria and plastids have an inner and an outer membrane. Plastids have an additional internal membrane system, the thylakoid system. Mitochondria occur in animal as well as in plant cells, plastids only in plants. Microbodies are surrounded by only one membrane. They harbour specific metabolic pathways and are, dependent on these, classified as peroxysomes or glyoxysomes.

It depicts several structures that we know by now: cell wall, nucleus, chloroplasts, vacuole and a thin rim of cytoplasm right at the cell wall. The picture does also show clearly recognisable substructures.

ORGANELLES

Organelles are membrane-surrounded functional unities of the cell. Mitochondria, the sites of respiration, and plastids, of which chloroplasts are the most striking members are maybe their best-known representatives.

NEW PREPARATION TECHNIQUES AND METHODS

Further results concerning intracellular structures could not be gained before the use of new preparation techniques (studies dating from the mid-fifties to the mid-sixties). Freeze etching proved to be an optimal method for the depiction of membrane surfaces. The resolution is high enough to recognise single macromolecules or macromolecular complexes, especially if they are arranged in the membrane in regular patterns as two-dimensional crystals .

The use of glutaraldehyde as a conserving agent allowed the depiction of tube-shaped structures, the microtubuli. They are part of the cytoskeleton and are essential for the movements of organelles and other intracellular particles, for the transport of chromosomes and as structural elements of flagelli (9+2 structure).

Among the peculiarities of a plant cell is the cell wall, whose layered structure could be confirmed by electron microscopical analysis. Layers occur also in collenchyma, stone or corked cells.

The use of highly specific markers or probes to localise certain molecules or classes of molecules is another important technique. It represents a bridge to the molecular level. To stress this further, it is also spoken of the molecular architecture of a cell.

The attempts to analyse electron microscopic pictures with the aim to reveal the three-dimensional shape of intracellular structures are independent of the mentioned methods. The huge amounts of data needed for this make computers indispensable. The biggest and up till now not completely solved problem is the structure recognition by the computer. A machine cannot decide, wether a particle of a certain grey colour is actually part of a structure recognisable by the observer or not. The single steps of a method for the construction of a three-dimensional model of dictyosomes developed by W. Meinhardt, J. Lockenhausen, W. J. Dallas and U. Kristen are explained in the following. An important step, the conversion of the electron microscopic picture into a sketch of the outline remains to be done by hand.

Cells Within Tissues

The cell wall is in many places perforated by pits. Plasma bridges, so-called plasmodesmata run through them and connect the protoplasts of neighbouring cells. The pits and plasmodesmata can occur in various shapes. They may have a very narrow lumen, may be branched and/or very numerous; they may also have very wide lumina and occur only sparsely. Especially striking pits can be found in the different cell types of the vascular tissues, they have already been mentioned. Examples are the pores in sieve plates or the pit areas in xylem elements. As we have been seen before, a number of cell and tissue types can be identified by means of light microscopy. But the use of the electron microscope allows a closer examination of the differences between certain cell types.

Light Microscopy

The research into all cell structures was and is dependent on microscopic techniques. Electron microscopy helps to study the substructures of organelles already known by light microscopy and extends even into the molecular range. But it can only be used for the examination of fixed and contrasted, in other words, dead cells and is therefore unsuitable for the depiction of intracellular movements or transport processes. These can be studied with advanced techniques of light microscopy, like UV-microscopy, AVEC-DIC and fluorescence microscopy with nonpoisonous dyes.

The use of UV microscopy is especially interesting when studying bundles of actin filaments that cannot be studied with electron microscopy due to the lack of a suitable contrasting agent. AVEC-DIC has become the method of choice for the research into intracellular movements and cytoskeletons. Fluorescence microscopy of living cells can be used to examine the membranous flow.

The complete repertoire of subcellular organelles and other inclusions was made visible by electron microscopy. But nothing living can be seen in an electron microscope. The nearest you get to the study of dynamic processes within the cell's plasma is the comparison of picture series. Several light microscopic techniques have been optimised, new extremely specific fluorescence dyes have been developed and the use of video cameras and computers optimised the documentation and evaluation of the pictures. Three of these methods shall be shortly reviewed here: UV-microscopy, AVEC-DIC and the use of new fluorochromes.

UV-Microscopy

UV-microscopy offers two principal advantages. On one hand UV light has a shorter wave length (less than 310 nm) than visible light (*lambda* = 400-800 nm), thereby reducing the resolution limit of ABBÉ to about 0.1 μm. On the other hand many subcellular structures absorb ultraviolet light due to their amount of proteins and nucleic acids and can thus be perceived even without additional contrasting.

Still, UV-microscopy did not gain general acceptance for a long time. The cause were the difficulties in the documentation of results. Not before the use of video cameras showed UV microscopy to advantage. Beside the cell nucleus it is possible to identify mitochondria, plastids, spherosomes, the Golgi apparatus and other structures without doubt. Additionally tubular elements of a diameter of 100 nm or more can be seen that could upon comparison with electron and fluorescence microscopic studies be shown to be parts of the endoplasmic reticulum. In the quiet plasma near the cell wall they can often be perceived as a polygonal network, while they are parallel to the current's direction in plasma currents.

Furthermore bundles of actin filaments can be depicted, that are impossible to detect in electron microscopic images due to the lack of a specific contrasting agent. In single UV microscopic images they can only hardly be distinguished from tubular ER. But their involvement in movements shows, that they are structures other than the ER. Additionally they can definitely be identified with special fluorochromes in fluorescence microscopy.

AVEC-DIC

N. S. Allen and R. D. Allen developed Allen's video-enhanced contrast, differential interference contrast technique named after them. It is based especially on the image evaluation of interference contrast microscopic takings. The pictures taken by a video camera can be digitalised and thus be computer-analysed. Upon subtraction can, for example, all identical structures in successive images be erased. As a result only all the moving particles are depicted. But it is also possible to subtract images of different depth of field planes, so that a background, which is in the way, can be filtered out. The technique has become the method of choice for the study of cellular or intracellular movements and the cytoskeleton during the last years.

N. S. Allen and R. D. Allen as well as I.K. Lichtscheidl and D.G. Weiss have used the technique for the examination of plasma movements in plant cells (*Chara, Allium cepa*, and others). Its advantages are only poorly reflected in the pictures shown here. Only in video films the whole range of movements within the plasma can be properly portrayed. By evaluation of the films the different speeds of movement of the different particles can be calculated. It can thus be shown that certain particles 'walk' along bundles of actin filaments and change their direction as soon as they reach a bundle with another polarity. It shows also that the ER functions as a mechanic obstacle for the movement of some particles.

New Fluorochromes

The lipophilic cation 3,3-dihexyloacarbocyanin-iodide (DiOC) proved to be the ideal marker for the ER and other membranous systems. Since the cells are not fixed during this procedure and since the fluorochrome does not - at least for some time - impair the cellular functions, is it well-suited for the study of the membrane flow in living cells. This is different with fluorescence-marked phalloidine, the poison of the amanita that turned out to be actin specific and can thus be used for the detection of actin filaments in fixed cells. Actin filaments are particularly common in the area around the nucleus and from there they expand across the whole protoplast. They are combined in thick bundles or cables near the nucleus, but the further they expand from the nucleus, the more do they split up.

REFERENCES

Bowen, W. R., *Experimental Cell Biology*, New York: The Macmillan Comp., 1969

Frey-Wyssling, A., K. Mühlethaler, *Ultrastructural plant cytology*, Amsterdam: Elsevier Publ. Comp, 1965

Hall, J. L. (ed.), *Electron microscopy and cytochemistry of plant cells*, Amsterdam: Elsevier/North Holland, 1978.

Kiermayer, O (ed.), "Cytomorphogenesis in plants", (Cell Biology Monographs 8) Wien - New York: Springer Verlag, 1981.

Lichtscheidl, I. K., D. G. Weiss, "Visualization of submicroscopic structures in the cytoplasm of Allium cepa inner epidermal cells by video-enhanced contrast light microscopy", *European Journal of Cell Biology*, 1988.

6

Plant Cell Nucleus

Nuclei are spherical or disc-shaped, although occasionally they may appear lobed. They vary in size in different species but are usually within the range 1–10 µm in diameter. Nuclei may be isolated by gently breaking cells in a suitable osmoticum, filtering the homogenate, for example through muslin, followed by low-speed centrifugation. Attempts are usually made to free the nuclei of contaminating cytoplasm, cell wall fragments and other organelles by centrifugation through sucrose. However, methods for obtaining high yields of intact, purified nuclei from higher plant cells are at present not very satisfactory. For certain purposes it is often easier to work with chromatin preparations which consist essentially of chromosomal and nuclear material obtained from broken nuclei.

The major part of the cellular DNA is present in the nucleus, arranged in a highly organised fashion within the chromosomes which at different times may show characteristic bands or loops. Chromosomes are composed of approximately equal amounts of DNA and basic proteins known as histones. In addition, much smaller quantities of RNA and acidic proteins are present which are probably of great importance in controlling gene expression. There is wide variation in the number of chromosomes observed within the nuclei of different species although in diploid cells of a single species the number is normally constant. During interphase the chromosomes are less highly condensed than at mitosis and are more difficult to detect. At this stage at least part of the DNA is being transcribed into RNA. It is possible to identify chromosomes on the basis of morphological features such as size, position of the centromere and the presence or absence of constrictions.

In addition, certain areas of heterochromatin, which appear more highly condensed and therefore take up more stain, may be distinguished from the remainder of the chromosome, termed euchromatin. In certain cell types, such as embryo suspensor cells, giant polytene chromosomes occur.

Recent developments have made it possible to identify individually each of the 46 chromosomes in man on the basis of characteristic banding patterns produced by certain staining procedures. These techniques and their application to the study of plant chromosomes have been reviewed by Vosa.

Organelles known as nucleoli become visible within the nucleus during interphase. These are very prominent spherical bodies with different staining properties to the remaining part of the nucleus. They arise in association with specific regions, the nucleolar organisers, of certain chromosomes. Nucleoli are not surrounded by a membrane but they do have a highly ordered structure. The central core consists of a region of fibrils 8–10 nm across surrounded by a region of dense granules 15–20 nm in diameter.

Mutants which lack nucleoli are unable to synthesise ribosomes and it is now known that the genes coding for ribosomal RNA normally occur at the nucleolar organiser region of chromosomes. Here the DNA is transcribed into ribosomal-RNA (rRNA) which is assembled to form ribosomes in the outer region of the nucleolus by combination with ribosomal proteins. The nucleolus is most prominent, therefore, when synthesising large numbers of ribosomes. This explains the old observation that there is a general correlation between nucleolar volume and the rate of protein synthesis of cells.

The fact that nuclei are surrounded by a membrane was deduced from the observation that they are selectively permeable to small molecules. The nature of the double membrane may be readily observed with the aid of the electron microscope. It is often interconnected with the cytoplasmic membrane system, the endoplasmic reticulum, and is perforated by a large number of pores. The apparent diameter of a nuclear pore is of the order of 100 nm. However, the observation that colloidal particles of much smaller size do not readily pass through these pores suggests that their effective size is substantially less, probably due to the presence of a protein-containing matrix within the pore.

NUCLEAR DNA CONTENT

The amount of DNA present in nuclei may be estimated by nuclear isolation followed by extraction and direct chemical measurement of DNA content. A more elegant approach is to carry out quantitative microspectrophotometry of individual nuclei following Feulgen staining. Table 1 compares the nuclear DNA content of a range of higher plants with the amounts of DNA found in lower organisms. Two conclusions can be drawn from such a comparison. Firstly, and perhaps not surprisingly, higher plants contain considerably more DNA than bacteria and fungi. For example, *Pisum sativum* contains 1,000 times as much DNA as *Escherischia coli* . Secondly, however, the amount of DNA present in different higher plants may vary by as much as 100 fold. Further comparison shows that there is little correlation between chromosome number and DNA content.

Table 1. DNA content of different organisms.

Genus and species	*DNA amount (picograms per 2C nucleus)*
Flowering plants	
Tradescantia ohioensis (4 ×)	89.0
Allium globosum	75.8
Tradescantia virginiana (4 ×)	70.0
Tulipa gesneriana	515
Scilla campanulata	46.0, 40.0
Tradescantia ohioensis	45.9
Allium karataviense	45.4
Scilla sibirica	44.5
Allium angulosum (4 ×)	41.2
Allium cepa	33.5
Vicia faba	28.7, 25.8, 26.8, 23.9
Chrysanthemum nipponicum	26.8
Allium jesolianum	24.1
Secale cereale	18.9, 17.7
Allium fuscum	18.4
Phalaris minor (4 ×)	17.1
Allium sibiricum	15.2
Lolium perenne	9.9
Helianthus annuus	9.8
Pisum sativum	9.1
Kalanchoë daigremontiana	6.7
Vicia angustifolia	6.1
Lycopersicum esculentum	3.9
Raphanus sativus	3.1
Cucurbita pepo	2.6
Lupinus albus	2.3
Linum usitatissimum	1.4
Microorganisms	
Fungus	*Haploid amount of DNA (picograms per cell)*
Aspergillus sojae	0.088

Aspergillus nidulans	0.044
Neurospora crassa	0.017
Bacterium	DNA per cell (picograms per cell)
Bacillus megatherium	0.070
Salmonella typhimurium	0.011
Escherichia coli	0.009
Bacteriophage	DNA per particle (picograms)
T_4	0.000025
T_5	0.000018
T_1	0.000007

One explanation of the increased DNA content of higher plants is that more complex organisms require a greater store of genetic material than bacteria and viruses. Although it is difficult to estimate precisely the extent of the increased coding requirement it is quite clear that this is in itself not a satisfactory explanation for the presence of such large amounts of DNA in some plants. There often exists a wide variation in the $_2$C DNA content of plant species within one: genus and it is unreasonable to conclude that small differences between related plants are a direct consequence of gross differences in nuclear DNA.

Furthermore, some plants contain 20 times as much DNA as mammals and yet the latter are normally accepted as comprising a more complex group of organisms. It therefore appears that the amount of nuclear DNA is not directly related to the complexity of the organism. It is also possible to argue on other grounds that the total amount of nuclear DNA is not required in order to code for additional proteins. Firstly, it is difficult to envisage functions to account for a 1,000 fold increase in the number of enzymes. Secondly, if each higher plant contained such a large number of essential genes it would, on numerical grounds, be prone to an unrealistically high incidence of mutation.

The size of the genome of plants with relatively low nuclear DNA contents is of the order of 2×10^{12} daltons. Given that the molecular weight of a gene required to code for a messenger-RNA (mRNA) molecule for an average-sized polypeptide is 10^6 daltons, this represents, in principle, 2×10^6 genes. A more realistic estimate would be that 1% of this DNA actually codes for proteins. Experimental evidence for this view and a discussion of the organisation, properties and possible function of the remaining DNA is given later.

It used to be widely proclaimed that when differences in ploidy are taken into account the amount of DNA in each nucleus of a single species is constant. The observation that the gene number remains constant, and the fact that many plant cells are totipotent, supports the view that differentiation involves the selective control of gene expression. However, the rule of DNA constancy is not inviolate. It is now known that specific regions of the genome may become selectively amplified at certain stages of development. This undoubtedly occurs to the genes for ribosomal RNA during oocyte development in Amphibia. Although there exist no established cases of ribosomal RNA gene amplification in higher plants, the satellite DNA present in melon may be under- or over-replicated in different tissues.

One very intriguing observation is that the amount of nuclear DNA in flax may become altered if the plants are subjected to certain regimes of nitrogen and phosphorus fertiliser. The alteration in nuclear DNA content is stable and inherited. Although this phenomenon is well established it is not known how widespread it may be and there is no satisfactory explanation to account for it.

FRACTIONATION AND PROPERTIES OF DNA

Although single-stranded DNA does occur naturally most DNA molecules consist of two anti-parallel chains of nucleotides. Each nucleotide consists of deoxyribose, a phosphate group and a base. The four bases that occur in all types of DNA are the purines adenine and guanine and the pyrimidines thymine and cytosine. Adenine is capable of forming hydrogen bonds with thymine, and guanine with cytosine. Sometimes modified bases are present in small amounts and in plant nuclear DNA a substantial proportion of the cytosine residues are replaced by 5-methyl-cytosine.

In double stranded DNA there is an equivalence between the amount of adenine and thymine and guanine and cytosine plus 5-methyl-cytosine. The two strands are joined together by hydrogen bonds between complementary pairs of bases to form a right handed helix. The bases are at right angles to the sugar residues, and project inwards, and the phosphate groups are on the outside. In the B form of DNA there are ten base pairs in one complete turn of the helix, which extend for 3.4 nm. The width of the helix is 1.8 nm. An alternative configuration is possible, the A form, where the bases are tilted and 11 pairs are accommodated per turn of the helix. Double stranded RNA and DNA-RNA hybrids adopt this latter configuration.

Many DNA molecules are known to be several millimetres in length. It is very difficult to determine the actual length of very large DNA molecules because they are extremely sensitive to shearing forces produced by various isolation procedures. It is known that the complete genome of *E. coli*, with a molecular weight of 2.9×10^9 daltons, consists of a single circular DNA molecule of length approximately 1 mM . The total amount of DNA in a higher plant cell is equivalent to a length of approximately 1 m, but it is not known how many molecules this comprises.

The base composition of DNA is normally expressed as the per cent G + C content. This value, which is characteristic of a given species, varies widely in different organisms and in bacteria may be anything from 30% to 70%. The base compositions of a number of plant DNAs are given in Table 2.

Buoyant Density Centrifugation

One of the most common methods of fractionating DNA is to exploit differences in density between different types of DNA in concentrated salt solutions. Caesium chloride is the most commonly used. The method involves prolonged high speed centrifugation of purified DNA in concentrated caesium chloride. The centrifugation produces a gradient of salt concentration in the centrifuge tube and the DNA migrates to occupy a position in the gradient which corresponds to its own buoyant density. A number of factors affect the buoyant density, such as the nature of the caesium salt, the presence of heavy metals or DNA-binding dyes, the pH and the temperature. Under constant conditions (usually 25°C in caesium chloride at neutral pH) the buoyant density of DNA is related to the GC content:

$$\%G + C \text{ content} = \frac{\text{buoyant density (g cm}^{-3}) - 1.660}{0.098} \times 100$$

DNAs with different base compositions can therefore be separated by this method.

Table 2. Base composition of plant DNA. (Moles per 100 moles.)

	Adenine	Thymine	Guanine	Cytosine	5-methyl cytosine	G+C
Ascomycetes						
Saccharomyces cerevisiae	*31.7*	*32.6*	*18.3*	*17.4*	—	*35.7*
Neurospora crassa	*23.0*	*23.3*	*27.1*	*26.6*	—	*53.7*
Green algae						
Chlamydomonas globosa	*19.0*	*19.5*	*30.3*	*28.2*	*2.75*	*61.25*
Spirogyra *sp.*	*30.7*	*30.4*	*19.2*	*19.8*	—	*39.0*

Gymnosperms						
Pinus sibirica	29.2	30.5	20.8	14.6	4.9	40.4
Ginkgo biloba	31.6	33.5	17.2	17.7	—	34.9
Angiosperms						
Daucus carota *(carrot)*	26.7	26.9	23.1	17.3	5.9	46.5
Zea mays *(maize)*	26.8	27.2	22.8	17.0	6.2	46.0
Triticum aestivum *(wheat)*	27.3	27.1	22.7	16.8	6.0	45.5
Cucurbita pepo *(pumpkin)*	30.2	29.0	21.0	16.1	3.7	40.8
Phaseolus vulgaris *(French bean)*	29.7	29.6	20.6	14.9	5.2	40.7
Allium cepa *(onion)*	31.8	31.3	18.4	12.8	5.4	36.3
Arachis hypogaea *(peanut)*	32.1	32.3	17.6	12.3	5.7	35.6
Gossypium hirsutum *(cotton)*	32.8	32.9	16.9	12.7	4.6	34.2

Most nuclear DNAs from higher plants have buoyant densities within the range 1.69–1.71 g cm^{-3}. However, the presence of 5-methyl-cytosine serves to reduce the density slightly, thereby giving rise to an under-estimate of the GC content. In general, 1% methylation decreases the buoyant density by 1 mg cm^{-3}. Certain sequences of bases may also distort the relationship between base composition and buoyant density. Furthermore, single-stranded DNA is denser than double-stranded DNA of similar base composition by approximately 0.015 g cm^{-3} and under alkaline conditions the density is increased by 0.06 g cm^{-3}. This is due partly to the fact that DNA becomes single-stranded under these conditions and also partly because the deprotonated adenine and thymine residues are neutralised by binding caesium ions.

Satellite DNA

In most plants certain regions of the genome are sufficiently different in base composition for them to occupy a position distinct from the main-band DNA. If they represent only a small percentage of the total DNA such sequences are difficult to detect but when present in large amounts they produce one or more peaks of satellite DNA. Melon DNA provides an example where the satellite is denser than the main-band but light satellites have also been observed, for example in flax. It has been assumed that the majority of satellite DNAs are nuclear in origin but recent studies with cucumber suggest that they may sometimes represent organelle DNA. Although they are widely distributed in plants no satellites have so far been found in monocotyledonous plants using caesium chloride fractionation. However similar components can be detected using other techniques. For example, substances such as heavy metals and dyes often selectively bind to certain sequences within the DNA. This alters their buoyant density and thus generates satellites. One common method is to fractionate DNA in caesium sulphate gradients in the presence of silver ions. Using this procedure, Ingle has demonstrated the presence of a satellite DNA in *Scilla* (a monocotyledon) although in caesium chloride no such satellite is observed.

Dissociation and Reassociation of DNA

When double-stranded DNA in solution is heated above a certain temperature the hydrogen bonds between complementary base pairs are broken and the two strands dissociate. The process, which can also be brought about by treatment of DNA with alkali, is commonly referred to as melting or denaturation. Strand separation is accompanied by an increase of approximately 35% in the absorbance of ultra-violet light. This hyperchromic effect provides a means of monitoring the dissociation of double-stranded DNA by measuring the increase in absorbance of a DNA solution at 260 nm as the temperature is raised. The melting curves of DNA from a variety of sources are compared in Fig. 1.

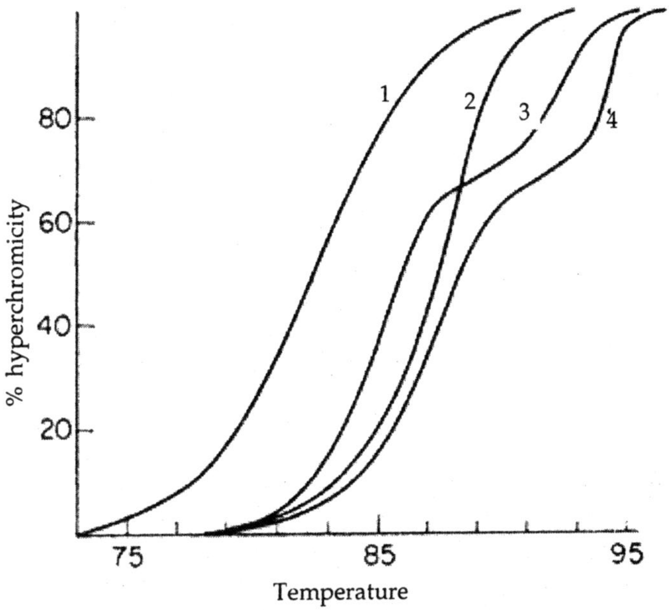

Figure 1. Melting curves of various DNA samples. (1) Musk melon main-band and (4) satellite DNA. (3) a 2:1 mixture of DNA from bacterium C and phage.

The temperature at which 50% of the DNA is melted, the Tm, depends upon a number of factors such as the salt concentration and pH. Under standard conditions, i.e. when the DNA is dissolved in standard saline citrate the base composition and the Tm are related in the following way:

$$\text{GC content (\%)} = (\text{Tm} - 69.3(\times 2.44$$

In viruses, the majority of sequences within the genome are similar in base composition and for a given DNA a sharp melting profile is observed with a characteristic Tm. In contrast, higher organism DNA often consists of distinct subfractions, each with a slightly different Tm. This produces a broad melting profile which represents a number of overlapping melting curves. With DNA samples from organisms that contain prominent satellite bands in caesium chloride the melting curve often has two or more phases. Even where no large satellites are present heterogeneity within the DNA can readily be detected by plotting the *rate of change* of absorbance at 260 nm of a DNA solution as a function of temperature.

If, following denaturation of DNA by alkali or by heating, the pH of the solution is restored to neutrality or the temperature reduced below the Tm, complementary strands of DNA show a marked tendency to recombine to form double-stranded molecules. This process of duplex formation is termed renaturation or reassociation. The rate of the reaction is governed by the concentration of DNA sequences, pH, temperature and salt concentration. Renaturation is normally carried out at moderately high salt concentrations at a pH close to neutrality and at a temperature approximately 25°C below the Tm.

Under these conditions, single-stranded DNA molecules collide at random. Most collisions occur in such a way that complementary sequences of bases are not in register and the strands do not reassociate. If, however, two strands approach in an anti-parallel configuration and potential regions of complementarity are in close proximity a nucleation event will take place and hydrogen bonds will form. Whether or not the strands remain together depends upon the nature of the neighbouring bases on either side of the nucleation site. If complementary sequences are arranged in the correct register along the length of the molecule a rapid zippering effect takes place to produce a stable double-stranded DNA molecule.

DNA renaturation can be followed by measuring any property that distinguishes single-stranded from double-stranded DNA. The decrease in absorbance at 260 nm that accompanies duplex formation is often exploited for this purpose but fractionation by hydroxyapatite is also commonly used. Under appropriate conditions only double-stranded DNA binds to hydroxyapatite; single strands pass straight through a column and double-stranded material may subsequently be eluted either by increasing the salt concentration or by raising the temperature above the Tm. This procedure has the advantage that the properties of the single-stranded and renatured fractions may subsequently be studied separately.

Repeated Sequence DNA

Information about the sequence content of DNA can be obtained by studying the renaturation rate. Double-stranded DNA is first sheared to lengths of a few hundred nucleotides by ultrasonication or by forcing through a narrow gauge needle. This is designed to liberate short sequences of DNA and to allow them subsequently to react independently of neighbouring sequences. The DNA is then denatured and allowed to renature at the desired concentration under controlled conditions. Renaturation is followed by monitoring the absorbance of the solution at 260 nm in a spectrophotometer cell maintained at the appropriate temperature, or alternatively by taking samples at intervals and fractionating them by hydroxyapatite chromatography. In some situations partially double-stranded segments of DNA are produced, which have single-stranded 'tails'. These can be removed by treating the reassociated DNA with SI nuclease, which is specific for single-stranded DNA, before hydroxyapatite fractionation.

The reassociation reaction follows second order kinetics, which means that it is governed by the concentration of the *reacting sequences.* This is determined by the DNA concentration and the sequence content. For example, synthetic poly-dA. dT renatures almost instantaneously, whereas at the sme concentration naturally occurring DNA takes much longer to renature. In general the renaturation rate is inversely proportional to the number of base pairs in the genome.

In principle, therefore, the number of 'genes' or sequences may be estimated by comparing the renaturation rate of an unknown DNA with that from an organism with a well characterised genome. The results of renaturation experiments are often expressed as a 'Cot' plot. In this form the extent of DNA renaturation is plotted as a function of the initial concentration of the DNA (in moles of nucleotides per litre) multiplied by the renaturation time (in seconds) plotted on a log scale. One advantage of this is that different types of DNA can easily be compared by their Cot 1/2 values. For a variety of microorganisms and viruses the Cot 1/2 is related to the genome size. However, this is not strictly true for eukaryotic DNA which behaves in a more complex way.

In bacteria and viruses most DNA sequences are represented only once in the genome but in higher organisms many sequences occur in multiple copies. For this reason a plot of the renaturation of higher plant DNA deviates markedly from a second order curve. A substantial proportion of the genome is so highly repetitive that it renatures even more rapidly than bacterial DNA. A second, intermediate, fraction contains families of repetitive sequences present in a lower frequency. Finally, there exists a unique fraction of DNA which essentially comprises sequences present in only one or a small number of copies in the genome.

Flavell *et al.*, have studied the reassociation kinetics of the DNA from 23 higher plant species, which vary in 2C nuclear DNA content between 1.7 and 98×10^{-12} g. For plants with smaller genomes the average amount of repetitve DNA is 62%. These results indicate that part of the variation in nuclear DNA mass can be accounted for by variation in the amount of repeated sequence DNA.

GENES FOR RIBOSOMAL-RNA

More is known about the properties, organisation and function of the genes for rRNA than any other part of the genome. These DNA segments direct the synthesis of rRNA molecules which form the structural 'backbone' of ribosomes, without which cells cannot function. It has been known for some time that they are associated with the nucleolus because certain mutants of the toad *Xenopus,* which lack nucleoli, are incapable of synthesizing rRNA. During development these mutants die because they are unable to synthezise ribosomes.

DNA-RNA Hybridisation

The genetic origin of any RNA molecule can be studied by DNA-RNA hybridisation. This technique is similar in principle to DNA-DNA reassociation. It depends upon the fact that any RNA molecule synthesized on a DNA template has a base sequence complementary to the DNA strand from which it was transcribed. Under suitable conditions, therefore, DNA can be induced to combine with complementary-RNA (cRNA) to produce an RNA-DNA double helix. This provides a means of locating and measuring the genes that code for a particular type of RNA.

One of the commonest procedures is to prepare pure DNA, denature it by heating or treatment with alkali, and to attach the single strands to a solid matrix such as a nitrocellulose filter. During subsequent treatments the immobilised DNA strands are prevented from reassociating but they are capable of forming hybrids when they are placed in a solution of nucleic acid containing complementary base sequences.

Using radioactive RNA the formation of DNA-RNA hybrids can be detected by measuring the amount of radioactivity that becomes associated with the filter. The reaction is affected by a number of factors such as RNA concentration, salt concentration, temperature, the presence of compounds such as formamide, which lowers the Tm, and the reaction time. With rRNA at a concentration of 2 μg cm^{-3} at 70°C in 6 × S.S.C. the reaction is virtually complete within a couple of hours (Fig. .2).

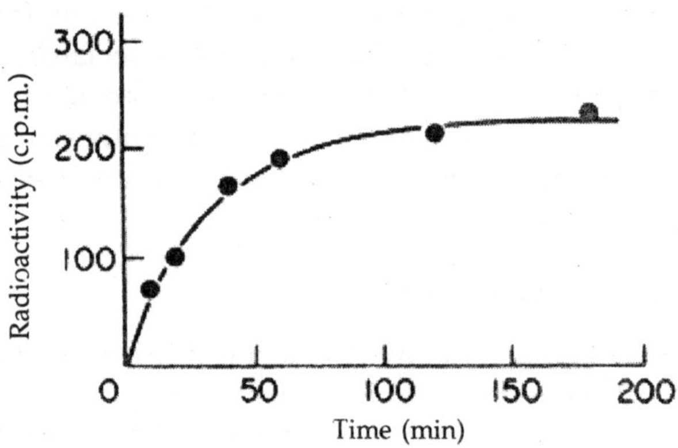

Figure 2. Saturation hybridisation of nuclear DNA from Phaseolus aureus with radioactive 25S rRNA.

After a suitable time the filters are washed, given a mild treatment with ribonuclease to remove any unhybridised RNA, and the amount of radioactive RNA remaining is measured in a liquid scintillation counter. If the amount of DNA initially applied to the filter is known then the extent of hybridisation can be calculated from the specific radioactivity of the RNA. The amount of DNA that hybridises to rRNA can either be calculated directly from the saturation plateau (Fig. 2) or, more accurately, from a double reciprocal plot of the data. In most plants between 0.05 and 1% of the DNA hybridises to rRNA. This represents from a few hundred to several thousand genes per diploid nucleus, depending on the species (Table 3).

The 258 and 18S rRNA molecules hybridise to separate sites on the DNA in accordance with the ratio of their molecular weights (1.3:0.7; approximately 2:1 by mass but 1:1 molar ratio).

Table 3. Number of ribosomal-RNA genes in different plants.

Plant	*% DNA hybridised*	*Number of genes per haploid complement*
Swisschard (*Beta vulgaris, v. cicla*)	0.20	1,150
Maize (*Zea mays*)	0.18	3,100
Wheat (*Triticum vulgare*)	0.092	2,100
Cucumber (*Cucumis sativus*)	0.96	4,400
Onion (*Allium cepa*)	0.090	6,650
Pea (*Pisum sativum*)	0.17	3,900
Artichoke (*Helianthus tuberosus*)	0.022	260

This indicates that there are equal numbers of genes for each type of rRNA. In cases where they do not hybridise in equimolar amounts this may be due to sequence homology between the two RNAs, or alternatively may arise by contamination of the RNA samples with other types of RNA.

Multiple genes for rRNA are a physiological necessity in order to provide sufficient sites of synthesis for rRNA molecules to meet the high metabolic demand for ribosomes. Plants often contain thousands of these genes although the number varies between varieties of the same species and between different species. Most of the available evidence suggests that all the copies are identical and this provides a clear example of repetitive DNA with a known function.

Their localisation within the nucleolus has been confirmed by *in situ* hybridisation of rRNA to cytological preparations.

In many eukaryotes ribosomal DNA occupies a separate position in caesium chloride gradients relative to main band DNA. However, since they represent 1%, or less, of the genome these genes are not easily detected by studying the distribution of the DNA. Nevertheless, they can readily be pinpointed by hybridising rRNA with separate fractions of DNA from the gradient. The fractionation of main band DNA from the rRNA genes of *Phaseolus aureus* is shown in Fig. 3.

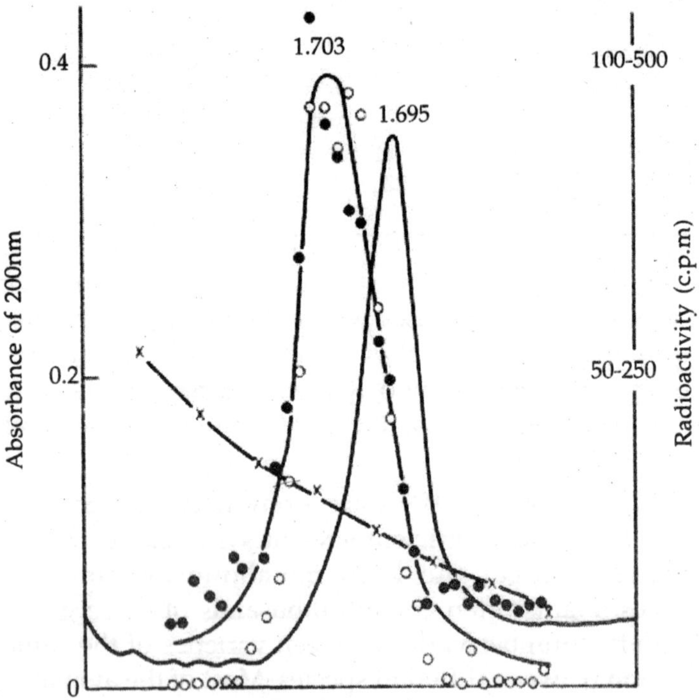

Figuer 3. Separation and detection of the DNA sequences coding for

In general, ribosomal DNA from plants has a higher density than main band DNA. This is consistent with the fact that in higher plants rRNA has a base composition in excess of 50% GC, whereas the bulk of nuclear DNA often has a lower GC content. In this sense rRNA genes represent a type of satellite DNA. However, in different plants the rRNA genes are often present in similar numbers and with a similar density irrespective of the presence or absence of large amounts of satellite DNA. It therefore appears that ribosomal DNA and other satellites may occasionally band together in caesium chloride but this is entirely fortuitous.

Electron Microscopy

Multiple rRNA genes therefore alternate in a precise arrangement with non-transcribed regions of the DNA. Hybridisation evidence shows that separate 'cistrons' for the 25s and 18s rRNA are present in the DNA, but there is no evidence for the existence of separate transcription units. In contrast, in *Acetabularia* each transcribed region of double-stranded DNA is approximately 2 μm long. This indicates that it codes for an RNA molecule with a molecular weight of approximately 2×10^6 daltons. This is the combined mass of the 25s and 18s rRNA molecules ($1.3 + 0.7 \times 10^6$ daltons). There is clear evidence (discussed in the section on RNA synthesis) that ribosomal RNA is first transcribed as a polycistronic precursor molecule.

In many plants this precursor has a molecular weight of approximately 2.4×10^6 daltons, and contains one sequence of 25s RNA plus one sequence of 18s RNA together with some extra sequences which are subsequently removed. In order to reconcile such a mode of synthesis with studies on the organisation of the rRNA genes it is necessary to propose that each 'matrix' unit consists of an 18s and 25s sequence and some transcribed 'spacer' DNA. Fortunately, in cases where the organisation of ribosomal DNA has been studied in detail, for example in *Xenopus*, evidence for precisely this type of arrangement has been obtained. Furthermore, recent studies with the electron microscope have provided evidence for a common organisation of the rRNA genes in all animals examined.

By carrying out partial denaturation studies it has proved possible to map certain areas rich in GC or AT base sequences. In addition, it is known that both the non-transcribed 'spacer' regions and the transcribed spacer can vary significantly in length and composition between species. Possible functions for such regions are discussed below. In contrast, however, the regions that code for rRNA are remarkably constant in composition. This probably reflects the importance of their role in providing structural RNA for ribosome formation.

As shown in Fig. 3, chloroplast rRNA also hybridises to nuclear DNA. This is a surprising result in view of the fact that chloroplasts contain their own DNA and synthesize their own rRNA. The location of these genes within the nucleus is not known, and at present there is no evidence that they are ever transcribed in the nucleus.

ORGANISATION AND FUNCTION OF THE GENOME

Repetitive DNA

Reassociation studies indicate that between 40% and 90% of the genome of higher plants is composed of repetitive DNA. This fraction is not homogeneous but consists of a number of types of sequence with different lengths and repetition frequencies. In hexaploid wheat, for example, 70%–80% of the genome is made up of families of sequences present in from 100 to 100,000 copies. The question of the composition, organisation and function of these sequences is still under investigation but at least three different classes can be distinguished. These are (a) clustered sequences of repetitive DNA which are transcribed into RNA, (b) clustered sequences which are apparently not transcribed, and (c) reverse-repeats distributed at intervals throughout a substantial part of the genome.

Transcription of repeated sequences

DNA-RNA hybridisation has shown that a small part of the repeated sequences represents genes for rRNA. These consist of regularly arranged alternating regions of transcribed and non-transcribed DNA. The transcribed sequences can be subdivided into those where the products are stable, the 18s and 25s rRNA; and those that have metabolically unstable products, the transcribed spacer DNA in the rRNA precursor. The role of the transcribed and nontranscribed spacer sequences has not yet been unequivocally established although obvious possibilities are that they may be involved in genetic organisation or regulation of transcription. Different groups of rRNA genes may sometimes be present on a number of chromosomes but as far as is known they are all practically identical.

Other families of repeated genes which have been shown to be transcribed include those for 5s RNA, tRNA and, at least in animals, the genes for histone proteins. What proportion of the genome do these account for? The actual value is higher than at first appears because hybridisation only detects one strand of the DNA and normally takes no account of either the transcribed or non-transcribed spacer. This means, for example, that a 0.5% hybridisation level for rRNA probably accounts for about 2.5% of the total DNA. Similar calculations for the 5s, tRNA and histone genes show that the total number of known sequences of this type only account for something like 5% of the genome.

Palindrome sequences

A small fraction of rapidly-renaturing DNA is known to be interspersed throughout a large part of the genome. The evidence for this comes from the 'zero-time binding' of DNA to hydroxyapatite. When DNA is denatured and immediately passed through hydroxyapatite a fraction of the DNA binds, suggesting that it is double-stranded. This happens even at very low Cot values where reassociation probably does not occur. The most convincing explanation of this behaviour is that it is due not to strand reassociation but to intra-strand 'snap-back'. This would occur if double-stranded DNA contained reverse-repeat sequences of the following type:

– – A – T – A – G – G – C – G – C – C – T – A – T – –

– – T – A – T – C – C – G – C – G – G – A – T – A – –

With this arrangement, termed a 'palindrome' sequence, the potential complementarity between adjoining regions of the same single-stranded DNA molecule results in rapid hairpin formation. In consequence any single-stranded DNA molecule will bind to hydroxyapatite providing it contains a double-stranded hairpin region. This provides a means of deducing the organisation of palindrome sequences in the genome. The amount of DNA that binds to hydroxyapatite at zero time is markedly dependent upon the fragment size of the DNA used. This suggests that the palindrome sequences are interspersed with other regions of DNA. It has been calculated that in wheat the average distance between these palindrome sequences is 8,500 nucleotide pairs. Between 4% and 10% of the DNA is palindromic and such sequences are spread throughout at least 40% of the genome.

There are several possible functions for such reverse-repeats. Any RNA molecule transcribed from DNA interspersed with sequences of this type would be expected to have double-stranded hairpin regions. These are known to occur in RNA but the proportion of palindromes in DNA is probably too high to be accounted for solely on this basis. One obvious property of palindromes that may have functional significance is that of symmetry. This may be important for the recognition of particular regions of double-stranded DNA by enzymes or regulatory elements.

Other types of repetitive DNA

Satellite DNAs also consist of repeated sequences. Although apparently homogeneous in caesium chloride the melon satellite has been shown to consist of at least two components, judged by melting experiments, caesium sulphate centrifugation and renaturation behaviour. Preliminary estimates are that the repeat lengths of the two sequences are respectively several hundred and a million or more nucleotide pairs. It is of interest to ask to what extent the copies within one family of repeated sequences diverge in sequence because this knowledge may provide clues about the origin and function of repetitive DNA.

Indirect information on this point comes from melting experiments carried out with native and renatured satellite DNA. When similar, but not identical, sequences renature the possibility exists for the formation of duplexes with a certain proportion of mis-matched bases. For every 1% mismatching in renatured DNA the Tm is decreased by 1.0 ± 0 3°C below that of native DNA. Judged by this criterion the sequences within the two types of melon satellite DNA are remarkably similar. Many animal satellite DNAs appear to have evolved by duplication of a rather short sequence of DNA followed by divergence from the basic sequence. They are believed not to be transcribed.

In contrast, the melon satellite DNAs appear to show very little sequence divergence and furthermore one component at least has a suprisingly large potential coding capacity. However, this may prove to be exceptional and further experimental work is necessary before it can be established whether or not plant satellites are more complex than those found in animals and whether they are ever transcribed into RNA.

The division of repetitive DNA into satellite and non-satellite components is probably an arbitrary distinction. All plants contain repeated sequence DNA irrespective of whether or not they possess satellites. This emphasises the considerable variation in properties of repeated DNA sequences. However, very little is known about their function. The fact that they sometimes form satellites suggests that they are clustered to some extent, otherwise they would remain attached to, and therefore band in association with, main-band DNA in caesium chloride. In the mouse, hybridisation to cytological preparations has shown that satellite sequences are concentrated around the centromeres of many of the chromosomes, but this is not always true of satellite DNA.

Where satellites are present the amounts may vary in different cell types of the same plant. Furthermore there is some evidence that at least part of the repeated DNA can be dispensed with altogether. For example, Beridze has investigated six species of *Phaseolus* and shown that three contain substantial amounts of satellite DNA whereas three others possess little if any satellite. Clearly, in these cases, the satellite is not essential and its presence or absence appears to have little effect on the plant phenotype. It is quite possible that at least some of the repetitive sequences in DNA are manifestations of evolutionary 'doodling'.

Unique DNA Sequences

One might conclude that the non-repetitive portion of the genome represents that fraction of the DNA that codes for proteins. However, taking into account the repetitive DNA fraction, the *amount* of unique sequence DNA in different species is found to vary at least ten-fold. One way to estimate the proportion of DNA transcribed is to carry out hybridisation experiments between mRNA and unique-sequence DNA. Due to the very high Cot values required the rate and extent of hybridisation using the conventional filtered technique are unsatisfactory.

An alterative approach is to carry out DNA renaturation at high concentrations, using only a trace of highly radioactive RNA for hybridisation. However, another technical difficulty is encountered in obtaining RNA preparations sufficiently enriched with mRNA and of sufficiently high specific radioactivity for this purpose and no estimates are yet available concerning the extent of transcription of the unique fraction of the plant genome.

RNA SYNTHESIS

Most types of RNA are single-stranded molecules transcribed by RNA polymerase enzymes from one of the two strands of a DNA template. RNA molecules are, therefore, the direct products of genes. The following section deals with the transcription of nuclear DNA and the processing and transport of RNA to the cytoplasm. The transcription of chloroplast and mitochondrial DNA and the special case of the synthesis of RNA on a RNA template by plant viruses are excluded.

Properties of RNA

Different types of RNA are characterised by their subcellular distribution and properties such as molecular weight and nucleotide composition. Approximately 70% of cellular RNA is contained in the ribosomes where it plays a structural role. Each 80s cytoplasmic ribosome contains three different types of ribosomal RNA (rRNA). These molecules are most commonly referred to by 's values' which are related to their rates of sedimentation during ultracentrifugation.

The larger ribosomal subunit contains two rRNA molecules of 25s and 5s and the smaller subunit contains a single rRNA of 18s. The approximate molecular weights are 1.3 × 10^6 daltons for 25s rRNA (about 3,900 nucleotides), 0.7 × 10^6 daltons for 18s rRNA (about 2,100 nucleotides) and 4 × 10^4 daltons for 5s RNA (about 120 nucleotides). The nucleotide compositions of a range of rRNAs are shown in Table 4. Transfer RNA (tRNA or 4s RNA) accounts for a further 25% of the total. In contrast to rRNA this fraction is heterogeneous. It consists of a range of similar but not identical molecules approximately 80 nucleotides long, each capable of being charged with a specific amino acid.

Table 4. Base composition of ribosomal-RNA and poly (A)-containing RNA from different organisms.

Moles per cent

Organism	rRNA	C	A	G	U	G + C
Yeast	*26s*	*19.2*	*26.4*	*28.4*	*26.0*	*47.6*
(Saccharomyces cerevisiae)	*18s*	*19.1*	*26.6*	*26.1*	*28.1*	*45.2*
Mushroom	*26s*	*20.1*	*27.7*	*27.8*	*24.3*	*47.9*
(Psalliota campestris)	*18s*	*18.0*	*24.9*	*30.1*	*27.1*	*48.1*
Artichoke	*26s*	*21.6*	*28.1*	*31.5*	*18.9*	*53.1*
(Helianthus tuberosus)	*18s*	*20.5*	*27.6*	*28.1*	*23.9*	*48.6*
Pea	*26s*	*21.8*	*24.9*	*32.2*	*21.1*	*54.0*
(Pisum satisum)	*18s*	*21.1*	*25.9*	*28.5*	*24.4*	*49.6*
Mung bean	*26s*	*23.0*	*24.5*	*32.4*	*20.1*	*55.4*
(Phaseolus aureus)	*18s*	*20.2*	*25.4*	*30.2*	*25.2*	*50.4*
Corn	*26s*	*28.0*	*21.3*	*33.9*	*16.9*	*61.9*
(Zea mays)	*18s*	*25.07*	*22.0*	*33.1*	*20.2*	*58.17*
Poly (A)-containing RNA						
French bean		*16.6*	*39.0*	*21.6*	*22.7*	*38.2*
(Phaseolus vulgaris)						
Sycamore						
(Acer pseudoplatanus)		*14.6*	*41.0*	*19.3*	*25.1*	*33.9*

By interacting with messenger-RNA appropriately charged tRNAs are responsible for aligning amino acids in the correct order during protein synthesis. In addition to the four major bases tRNA contains a very high proportion of modified or unusual bases. The most interesting type of

RNA, mRNA, is present in small amounts and constitutes less than 5% of the total. It is this fraction that directs the synthesis of specific proteins by the ribosomes. The length and nucleotide sequence of mRNAs are related to the properties of the proteins for which they code. As most plant cells synthesize a variety of different proteins the mRNA fraction generally consists of a range of molecules of different size and nucleotide composition.

Ribosomal RNA Synthesis

Ribosomal RNA Precursor

The DNA that codes for rRNA is known to be associated with nucleoli and the transcription of nucleolar DNA has been studied with the electron microscope. More detailed information has been obtained by the labelling of intact seedlings, plant segments, or cultured cells, with [^{32} P]-phosphate or [^{3} H]-uridine for various times followed by the extraction and characterisation of the rapidly-labelled RNA. The mechanism of rRNA synthesis is fairly well understood and a generalised scheme outlining the main features of the process is given in Fig. 4.

The first stable gene product which can be detected is an rRNA precursor molecule. The molecular weight of this RNA varies in different plants but is generally within the range 2.2–2.6 $\times$ 10^6 daltons. The precursor undergoes considerable post-transcriptional modification in the nucleolus before it is converted into 'mature' rRNA and enters the cytoplasm. Shortly after synthesis the molecule becomes methylated at selected sites. This can be demonstrated by incubating plants in methyl-labelled methionine for a short time. The methyl groups become incorporated into the rRNA precursor *via* s-adenosyl methionine. Ribosomal-RNA is known to contain 2'-methyl ribose at certain positions. Only a small number of the residues are modified in this way and it is probable that methylation is completed during the rRNA precursor stage.

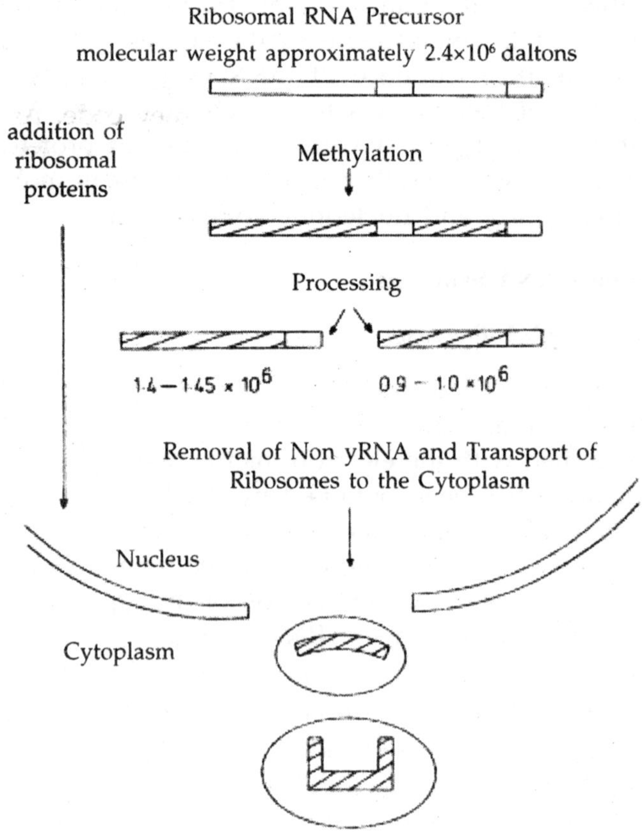

Figure 4. The synthesis and processing of the precursor to rRNA.

The precursor molecules correspond to the lateral fibrils attached to the matrix units of active nucleolar DNA. Each molecule contains one sequence of 25s rRNA, one sequence of 18s together with some non-rRNA. The evidence for this is as follows. Firstly, although it is rapidly labelled, the precursor does not accumulate in large amounts. This suggests, therefore, that it must either be rapidly degraded or converted to some other molecular form. This latter explanation would account for the fact that radioactivity appears in mature rRNA slightly later than in the precursor. Secondly, the nucleotide composition of the rRNA-precursor is similar to that of rRNA. Thirdly, competition hybridisation experiments have shown that both unlabelled 25s and 18s rRNA compete with radioactive precursor-RNA for sites in the DNA, suggesting that they share common sequences (Fig. 5).

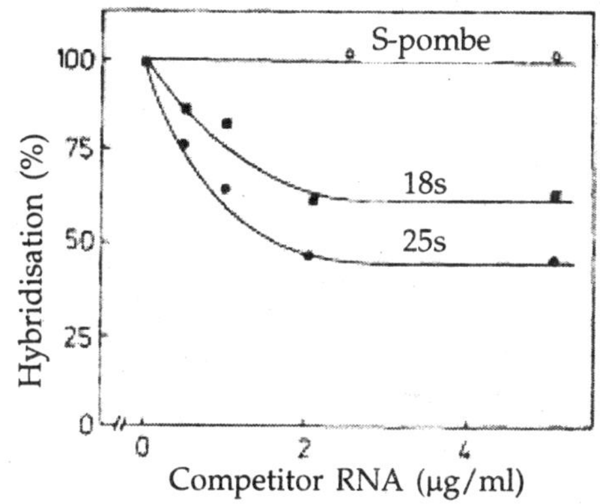

Figure 5. Competition hybridisation between unlabelled rRNA and radioactive rRNA precursor for similar sites in P. aureus DNA.

Finally, partial digestion of radioactive 25S, 18s and precursor RNA by T1 ribonuclease, followed by two-dimensional fractionation of the fragments shows that the 'fingerprint' pattern of the precursor is similar to that expected of a mixture of 25s and 18s rRNA.

Processing of the Precursor

In addition to becoming methylated, and while still in the nucleolus, the precursor becomes associated with protein molecules and is processed by nucleolytic ensymes which remove the excess non-rRNA in stages to produce mature rRNA in ribosomal subunits ready for transport to the cytoplasm. No immediate precursor to 25s rRNA is detectable in rye embryos. Furthermore the 1.0×10^6 daltons RNA is not always detected wheras certain additional RNAs are sometimes observed. For example, after cycloheximide treatment of cultured parsley cells, which slows down rRNA processing, previously undetected RNAs with molecular weights of 2.0 and 0.9×10^6 daltons were found.

This suggests that if certain processing events occur very rapidly they may normally go undetected. It is very probable that the complete sequence of processing steps will only be worked out when it becomes possible to study accurately RNA metabolism in isolated nucleoli. Apparent differences between closely related species may be explained by assuming that certain critical stages of processing occur at different rates. In addition, it should be realised that the processing enzymes are functioning at different, possibly distant, sites within the same molecule. There seems no compelling reason to expect these modifications always to occur in the same chronological order and, if they do not, variations in the pattern of processing intermediates might be expected under altered conditions or in different species.

The function of the excess RNA in the precursor is not known. It is too variable in size in different organisms to be a mRNA for the ribosomal proteins and in any case it appears to be degraded very rapidly after synthesis. It may play some role in the regulation of transcription but more probably it is necessary for ensuring that rRNA adopts the correct secondary and tertiary structure required for methylation and for the addition of proteins during ribosome assembly.

It is pertinent to ask whether the multi-stage mechanism that operates for rRNA production is unique or whether it represents a general principle common to the synthesis of all types of RNA. Precursors to tRNA and 5s RNA have been identified and characterised in bacteria and animals and precursors of mRNA have also been detected, although there is uncertainty about the actual details of processing. Although RNA metabolism has not been so intensively studied in plants the indications are that precursors to 5s RNA, tRNA and mRNA do occur. When considering the question of the regulation of RNA metabolism, therefore, it is necessary to take account of post-transcriptional modifications as possible control points.

Messenger-RNA Synthesis and Metabolism

The question of mRNA synthesis has proved extremely difficult to investigate due largely to problems of identification and purification of mRNA. In the past, rather imprecise criteria for mRNA identification, such as size heterogeneity, similarity in nucleotide composition to DNA, rapidity of labelling and association with polyribosomes, were adopted. More recently, two additional approaches have become available. Firstly, the development of an *in vitro* protein synthesizing system has made it possible to test the capacity of an RNA fraction to direct the synthesis of specific proteins. Secondly, the discovery that many mRNAs contain sequences of poly (adenylic) acid (poly (A)) has provided a simple approach to the purification of mRNA.

Poly (A)-Containing RNA

It has been known for some considerable time that a small fraction of total plant RNA has a very high AMP content. This fraction, often referred to as DNA-like RNA or D-RNA, is considered to consist, at least in part, of mRNA and is partially purified by chromatography on columns of methylated albumin adsorbed to kieselguhr. More recent work with animal mRNA has shown that the high AMP content is due to the presence of regions of poly (A) in the mRNA molecules. These sequences of poly (A) are generally 50 to 250 nucleotides in length.

In animals they are attached to the 3' end of the mRNA molecules *after* transcription although there is some evidence that shorter regions of poly (A) may also be introduced into RNA as part of the transcription process. Poly (A)-containing RNA has been identified in a variety of animals and plants although it has not been found in higher plant chloroplasts. It is apparently present in mitochondria from some organisms. This may reflect true differences between organisms or alternatively may be a question of the experimental technique. For example, contrary to earlier reports that poly (A) is absent from mRNA from bacteria, it has recently been shown to be present at a low level in *E. coli*.

The function of the poly (A) sequences is not known but their presence can be used to purify mRNA. There are a number of methods for doing this. Poly (A) will bind to nitrocellulose or cellulose at high salt concentrations. RNA without poly (A) can be washed off and the poly (A)-containing RNA is then eluted at low ionic strength. However, cellulose and nitrocellulose do not necessarily retain all the poly (A)-containing RNA. A more specific type of binding is achieved using columns of oligo (deoxythymidylic) acid attached to cellulose or poly (uridylic) acid attached to sepharose. Both these polymers rapidly form stable nucleic acid hybrids with the poly (A) sequence attached to the mRNA. Other types of RNA pass straight through the column, although it is sometimes necessary to wash through extensively with a buffer. The poly (A)-containing RNA is only released after lowering the salt concentration or raising the temperature to cause melting of the hybrid between poly (A) and either oligo (dT) or poly (U).

There is considerable evidence that poly (A)-containing RNA from plants is mRNA. It is associated with polyribosomes and its synthesis is inhibited by the antibiotic *a* -amanitin which prevents transcription of non-rRNA in the nucleus. When purified from cells synthesizing a large number of proteins of different sizes poly (A)-containing RNA is polydisperse, but when extracted from soybean root nodules, which synthesize substantial amounts of leghaemoglobin, two peaks of RNA are detected, each presumably consisting of similar or identical RNA molecules.

These RNA fractions stimulate protein synthesis by a cell-free system from wheat germ. Furthermore, leghaemoglobin has been positively identified among the products of *in vitro* synthesis. Supporting evidence that poly (A)-containing RNA stimulates protein synthesis comes from a large number of studies and the small number of individual proteins identified as *in vitro* products is growing steadily. However, it should be stressed that not all mRNA molecules contain poly (A). In animals, for example, about 70% of mRNA is estimated to contain poly (A) whereas only about 40% of polydisperse RNA from polyribosomes of cultured sycamore cells can be purified by oligo (dT)-cellulose. The functional significance of the two mRNA populations is not clear.

Postranscriptional Modification of mRNA

Poly (A)-containing RNA is synthesized in the nuclei of higher plant cells. Total rapidly-labelled RNA from nuclei consists of the rRNA precursors and processing intermediates together with polydisperse RNA. Approximately 15% of this RNA is retained by oligo (dT)-cellulose and by this criterion is judged to contain poly (A) sequences. It is polydisperse in size and has a high AMP content (Table 4). There is little direct evidence concerning the mechanism of poly (A) addition to nuclear polydisperse RNA in plants. It is assumed to be added to the 3' end of the molecules by the enzyme ATP-polynucleotidylexotransferase after transcription by RNA polymerase. The poly (A) sequences in higher plant RNA have been estimated to range from 50–250 nucleotides in length. However, plant DNA appears to contain short sequences of AT base pairs which could give rise to short poly (A) regions by transcription and there is evidence that similar regions are transcribed in slime moulds.

Comparison of the nuclear and cytoplasmic poly (A)-containing RNA by gel electrophoresis suggests that nuclear RNA is reduced in size before entering the cytoplasm. This suggests that there may be a processing mechanism analogous to that which operates for rRNA synthesis. In slime moulds the nuclear precursor appears to be approximately 20% larger than cytoplasmic mRNA. In higher plants the average molecular weight of the nuclear poly (A)-containing RNA is 1×10^6 daltons. This is in contrast to the giant precursors to mRNA earlier claimed to occur in animal cells. Very little is known about the processing mechanism, but Firtel and Lodish have shown that the nuclear precursor contains a sequence of about 300 nucleotides near the 5' end of the molecule which is lost before the RNA enters the cytoplasm. This is of considerable interest because this region appears to be transcribed from *repetitive* DNA. It therefore seems that some of the repeated DNA sequences interspersed between nonrepetitive DNA are transcribed and may play a regulatory role in processing or transport of mRNA.

NUCLEIC ACID-PROTEIN INTERACTIONS

Histones

Chromosomal DNA is invariably associated with several different types of proteins (Table 5). These include histones which occur in most eukaryotes.

There are five main histone proteins designated F_1, F_{2b}, F_{2a1}, F_{2a2} and F_3, with molecular weights from 11,000 to

22,000 daltons. They are strongly basic and generally have a high content of arginine and lysine. The basic amino acids are often clustered towards either the N or C terminus of the proteins suggesting that these regions interact directly with the DNA. Stedman and Stedman proposed that histones are responsible in some way for the regulation of gene expression but it now seems that they lack the specificity for recognition of different gene sequences and it is more likely that they fulfil a more general role in the organisation of chromosomes.

Comparison of the amino acid sequences of histones from different species suggests that they have been highly conserved during evolution. For example, the sequences of F_{2a1} from calf, pig and rat are identical and differ only in two substitutions from F_{2a1} from pea. This suggests these proteins evolved to fulfil an essential role with strict requirements for a precise sequence, although other histones, notably F_{2b} and F_{2a2} show greater variability. Histones frequently undergo modifications such as methylation, acetylation or phosphorylation.

Table 5. Chemical composition of varied chromatins

	Content, relative to DNA, of			
Source of chromatin	*DNA*	*Histone*	*Non-histone protein*	*RNA*
Pea embryonic axis	1.00	1.03	0.29	0.26
Pea vegetative bud	1.00	1.30	0.10	0.11
Pea growing cotyledon	1.00	0.76	0.36	0.13
Rat liver	1.00	1.00	0.67	0.043
Rat ascites tumour	1.00	1.16	1.00	0.13
Human HeLa cells	1.00	1.02	0.71	0.09
Cow thymus	1.00	1.14	0.33	0.07
Sea urchin blastula	1.00	1.04	0.48	0.039
Sea urchin pluteus	1.00	0.86	1.04	0.078

These are normally reversible and would be expected to alter the association of histones with DNA. In general these modifications occur at about the time of DNA synthesis although they may also herald an increase in RNA synthesis. During the cell cycle the synthesis of histones is closely linked to that of DNA and phosphorylation of F_1 has been shown to precede cell division. Basic non-histone proteins and acidic proteins also occur in association with DNA. These form a more heterogeneous group than the histones and include components of different molecular weight, charge and tissue distribution. They are less stable metabolically than histones and do not behave as a single class of proteins. Several lines of evidence suggest that this group may include regulatory elements.

Subunit structure of chromatin and chromosome organisation

The double helical structure of DNA is widely known and it is equally widely recognised that a further degree of organisation is responsible for the folding of 1 m or more of DNA into each set of chromosomes. Early observations on the structure of chromosomes indicated that DNA is organised in fibres of the order of 10 nm thick and it was suggested that the double helix, in association with proteins, is supercoiled. In addition, X-ray studies suggested that there is a repeating structure in chromatin at intervals of approximately 10 nm and more recently a linear arrangement of beads of the order of 7 nm diameter has been observed in chromatin with the aid of the electron microscope. Further evidence for a regular arrangement of DNA in chromatin comes from the observation that following nuclease digestion reproducible fragments of DNA approximately 205 base pairs long are produced. Kornberg has suggested a repeating subunit structure of approximately 200 DNA base pairs associated with 2 molecules of each of the four major histones (F_{2b}, F_{2a1}, F_{2a2} and F_3) and one molecule of F_1. In this model the DNA double helix is envisaged to be folded to approximately one seventh its length and Crick and Klug have suggested that this could occur by 'kinking' of the DNA helix at intervals of perhaps 20 base pairs. Bradbury and his colleagues have proposed that repeating units of DNA approximately 200 base pairs long are coiled on the outside of aggregates of the four major histones which associate with each other by apolar interactions, leaving their basic regions free to associate with the DNA. Histone F_1, which is believed not to be directly involved in the subunit, may perform a cross-linking function. Only about 87% of the DNA in chromatin is protected against deoxyribonuclease digestion and it is possible that not all the DNA participates in formation of this type of subunit.

Further insight into the functional organisation in chromatin comes from studies on the giant chromosomes from certain insects. These chromosomes consist of a series of densely coiled bands connected by less dense regions. These are much larger than the repeating unit of 200 DNA base pairs. Each band plus interband region is considered to correspond to a gene because there is good numerical agreement between their number and the number of complementation groups determined by breeding experiments.

However, calculations show that only a fraction of the DNA in each band plus interband region is necessary in order to code for a protein and the question of the function and organisation of the DNA remains unresolved. Regions of the DNA being actively transcribed appear as chromosome puffs composed of DNA, RNA and protein. Crick has suggested that the interband regions contain DNA that is transcribed and that each band consists of regulatory DNA sequences. In contrast, Paul has proposed a different type of arrangement.

Different regions of chromosomes often contain chromatin condensed to different extents. Euchromatin appears to be loosely packed. It is known to contain both unique and repeated sequences and is actively transcribed. Heterochromatin is more tightly packed and consists of two types. Constitutive heterochromatin contains sequences that are never transcribed and includes repeated sequences. In some species satellite DNAs are known to be clustered in heterochromatic regions near the centromeres. Finally, facultative heterochromatin constitutes those regions of the chromosome that sometimes appear as euchromatin.

Regulation of transcription

Thus there is evidence for the coiling of DNA and formation *of* regular DNA-protein complexes which may be further folded into larger units. Different regions of the chromosome are folded to different extents and this presumably affects the availability of DNA for transcription. Comparison of the extent of transcription of pure DNA and isolated chromatin by added bacterial RNA polymerase suggests that in animals only about 10% of the genome is available for RNA synthesis. Hybridisation experiments suggest that the RNA synthesized from isolated animal chromatin is tissue specific and resembles in sequence the type of RNA synthesized *in vivo*.

It is probable that the non-histone proteins are more important in determining which segments of the genome are repressed but further characterisation of the properties and mechanism of action of these proteins is necessary before their function becomes clear. It has also been suggested that chromosomal RNA may be important in regulating gene expression, possibly in conjunction with repeated sequences in DNA. There is undoubtedly RNA associated with chromatin, including growing chains of polydisperse- RNA and rRNA-precursor, but the existence of an additional type of regulator RNA has been proposed. However, at the present time there is a certain amount of controversy concerning the characterisation and properties of chromosomal RNA. As yet there is no evidence that this RNA regulates transcription.

In addition to repression of the genome by interaction with proteins, template selection may also be controlled by RNA polymerase enzymes and by other protein factors that interact with the polymerases to regulate transcription. As far as studies on plant chromatin are concerned *in vitro* RNA synthesis has been achieved using the endogenous RNA polymerases that remain attached to the isolated chromatin. However, there have been few attempts to characterize the RNA products in detail. There are several distinct RNA polymerases with different properties.

These enzymes are probably responsible for transcribing different regions of the genome. For example, RNA polymerase I, which is insensitive to the antibiotic *a* -amanitin, is located within the nucleolus. This enzyme appears to be responsible for transcription of the rRNA genes. The activity of this enzyme and the rate of RNA production are significantly enhanced in tissue treated with auxin. Enzymes sensitive to *a* -amanitin are present in the nucleoplasm and presumably are responsible for transcribing other parts of the genome. These polymerases may also be affected by hormones *via* the mediation of hormone-binding factors. In addition, evidence has been obtained for the existence of other protein factors which govern the activity of RNA polymerases by stimulating initiation of RNA synthesis.

REFERENCES

Bryant J.A., "Nuclear DNA", In *Molecular Aspects of Gene Expression in Plants* (ed. J.A. Bryant), pp. 1-51. Academic Press, London, 1976.

Grierson D., "RNA structure and metabolism", In *Molecular Aspects of Gene Expression in Plants* (ed. J.A. Bryant). pp. 53-108. Academic Press, London, 1976.

Kornberg A., *DNA Synthesis*, W.H. Freeman & Co., San Francisco, 1974.

Lewin B., "Gene Expression", Volume 2, *Eukaryotic Chromosomes*, J. Wiley & Sons Ltd., London, 1974.

Parish J.H., *Principles and Practice of Experiments with Nucleic Acids*, Longman, London, 1972.

Rae P.M.M., "The distribution of repetitive DNA sequences in chromosomes", In *Advances in Cell Biology and Molecular Biology* (ed. E.J. Dupraw), Vol. 2, pp. 109-49. Academic Press, New York, 1972.

Stewart P.R. & Letham D.S. (eds.), *The Ribonucleic Acids*, Springer-Verlag, Berlin, 1973.

7

Plant Mitochondria

Mitochondria are the sites of non-photosynthetic energy transduction in eukaryotic cells which, carry out aerobic metabolism. Energy transduction includes those processes by which the chemical potential energy of organic substrates is transformed into a readily mobilised form, adenosine-5'-triphosphate (ATP). Organic substrates are oxidised and the free energy of oxidation is conserved by processes which are common to all mitochondria, regardless of source. Thus, with regard to the oxidative and phosphorylative processes, information obtained from studies in animal mitochondria is applicable to plant mitochondria.

Notable differences between plant mitochondria and animal mitochondria do occur, although these differences do not contradict the basic similarities in the mechanism of energy transduction. For example, plant mitochondria possess external reduced nicotinamide adenine dinucleotide (NADH) dehydrogenases which oxidise exogenous NADH; mitochondria from animal sources lack this capability. Mitochondria from many plant sources are relatively insensitive to cyanide inhibition, a feature not found in animal mitochondria. On the other hand, the *b*-oxidation pathway of fatty acids is located in animal mitochondria, whereas in plants, the enzymes of fatty acid oxidation occur in the glyoxysomes.

MORPHOLOGY

Morphology in Situ

Mitochondria in living cells are highly pleomorphic, as shown by phase contrast microcinematography by Hongladarom *et al.* Pleomorphism is reflected also in thin section electron microscopy, in which mitochondria appear as roughly circular profiles as well as highly elongated or irregular cross sections. The circular sections may represent transverse or oblique sections through an otherwise elongated organelle. The diameter of the elongated mitochondrion appears to be about 0.4 to 0.5 μm, while the length may be several micrometers long. Although rods or apparent spheres are the most common profiles seen, sections derived from branched or cup shaped organelles have also been discerned. The recent analysis of serial sections of yeast cells by Hoffman and Avers, which showed that yeast contains a single, giant, branched mitochondrion, suggests that the irregular cross sections of mitochondria of other cells might also be sections of a single branched organelle.

The mitochondrion consists of a double membrane system with an inner convoluted membrane enclosing the matrix, and surrounded by a smooth outer membrane. High resolution electron micrographs of material fixed with glutaraldehyde and post-fixed with osmium tetroxide show the tripartite nature of both the inner and outer membranes. Each membrane has a thickness of approximately 9 nm.

Morphology of Isolated Mitochondria

Electron micrographs of isolated mitochomdria show circular cross sections, presumably reflecting a spherical shape when released from their cellular environment. The electron micrographs of the intact isolated mitochondrion show clearly the two membrane systems, as well as the tripartite organisation of each membrane. The fine structure of isolated mitochondria is highly dependent upon the osmolarity of the suspending medium. When mitochondria are suspended in 0.3 to 0.4 M sucrose or mannitol, the matrix appears contracted and electron dense, but when suspended in 0.2 M sucrose, the matrix appears more expnaded and less electron dense, and resembles that of mitochondria seen *in sity* . The dense matrix of mitochondria suspended in 0.3 to 0.4 M sucrose or mannitol is due to the hypertonicity of the suspending medium. Since the inner membrane is generally regarded as the osmotic barrier, the dense nature of the matrix reflects a water loss, which is reversible when the organelles are suspended in 0.2 M sucrose.

Negatively-stained water-lysed mitochondria show that the inner membranes have the characteristic stalked particles similar to those reported for mammalian mitochondrial membranes. The particles have a headpiece with a diameter of 10 nm, attached to a stalk 3.5 to 4.5 nm wide and 4.5 nm long, These resemble the particles identified with ATPase function in heart mitochondria.

Techniques of Isolation and Purification

Mitochondria have been obtained from a large number of plant sources including roots, storage tissue, stems and photosynthetic tissues. The usual problems of isolation, regardless of the source, are (a) the rupture of a rather rigid cell wall and (b) the prevention of damage to organelles through the release of intracellular, particularly vacuolar, contents. Ikuma listed a number of conditions for successful isolation of tightly-coupled mitochondria. These include (a) gentle tissue disruption, (b) rigorous exclusion of contaminating particles and (c) the use of a buffered grinding medium isotonic with mitochondria and containing a variety of protective reagents. Most investigators employ some device to reduce quickly the tissue to a coarse slurry, which is passed through a cloth filter to remove large debris.

The fraction which sediments between 1,000 *g* and 10,000 *g* is collected as the mitochondrial fraction. This fraction will oxidise all the intermediates of the Krebs tricarboxylic acid cycle, exhibit respiratory control and yield ADP to O ratios approaching the theoretical value for the substrate used. The mitochondrial fraction can be further purified by density gradient centrifugation. This may be done in discontinuous sucrose gradients or Dextran-40 gradients. Mitochondria form a band at the interface between 1.2 and 1.5 M sucrose.

This is recovered and diluted slowly to 0.3 M sucrose. This procedure yields mitochondria with intact outer and inner membranes as shown by electron microscopy. The integrity of the outer membrane is also shown enzymatically by the inability to reduce exogenous cytochrome *c* with NADH or succinate as substrates, unless the mitochondria have been subjected to mild osmotic shock which renders the outer membrane permeable to high molecular weight solutes.

During the disruption of cells and throughout the isolation procedures, a number of protective reagents must be present. Inclusion of sodium ethylenediamine-tetracetate (EDTA) in the isolation medium has been shown to give mitochondria with high respiration rates. EDTA probably removes cationic inhibitors although the specific cation complexed is unknown.

Many plant cells release phenolic compounds when ruptured. These phenolic compounds are oxidised in the presence of air and form polymers which are inhibitory to mitochondrial respiration and coupled phosphorylation. To prevent the damaging effects of phenolic compounds, a variety of reagents have been used successfully. Polyvinyl pyrrolidone is a competitive inhibitor of purified phenolase, and has been used extensively as a protecting agent for mitochondrial isolation. Other reagents include morpholinopropane sulphonate, cysteine, and sodium metabisulphite. Morpholinopropane sulphonate is thought to form complexes with phenolic compounds, while sulphydryl reagents inhibit phenoloxidases.

Free fatty acids are known to uncouple oxidative phosphorylation from electron transport. The uncoupling activity of fatty acids is reversed by the addition of bovine serum albumin. Bovine serum albumin also reverses the uncoupling activity of many other uncoupling agents, such as nitro-and halo-substituted phenols, dicumarol, and carbonyl-cyanide *m*-chloro-phenylhydrazone in rat liver mitochondria. Dalgarno and Birt showed that free fatty acids were present in mitochondrial preparations from carrot root tissue.

These included oleic, stearic, palmitic and some short chain fatty acids, as well as polyunsaturated C_{18} acids. Mitochondria isolated from such tissues were uncoupled as shown by a P/O ratio less than 0.1. When bovine serum albumin was included in the isolation medium, mitochondria became well coupled, with a P/O ratio greater than 1.6 with succinate as substrate. As a matter of routine, most isolation procedures include 0.1% (w/v) of bovine serum albumin (Cohn Fraction V, low in free fatty acids). Bovine serum albumin binds fatty acids, and other lipophilic uncoupling agents, but the nature of the binding is not clear.

Isolation from Green Tissues

Mitochondria isolated from photosynthetic tissues are rarely free from chloroplasts or chloroplast fragments. Rocha and Ting subjected spinach leaf material to linear sucrose gradients (40 to 80% w/v) and obtained fractions after equilibrium. They found, nonetheless, that the mitochondrial fraction was contaminated with 13% intact chloroplasts and 6% broken chloroplasts. The degree of contamination was estimated from the activities of characteristic marker enzymes. Malate dehydrogenase and cytochrome *c* oxidase served as mitochondrial markers, while chlorophyll content and triose-phosphate dehydrogenase were chloroplast markers.

MITOCHONDRIAL MEMBRANES

Structure of Membranes

Electron micrographs of plant mitochondria show clearly the tripartite nature of both the outer and inner membranes. This may be interpreted as a lipid bilayer with the hydrophobic fatty acid chains oriented toward the interior of the bilayer. The lipids of mitochondrial membranes are largely phospholipids. It is the current view that phospholipid bilayers are highly dynamic, with a high degree of fluidity in the fatty acid region, as well as high lateral mobility of the phospholipids in the plane of the membrane.

The membrane proteins may form loose interactions with the lipid bilayer, or may be very tightly associated with the membrane. The loosely associated proteins most likely have exposed hydrophilic side chains and are easily extracted from membranes. The proteins tightly associated with membranes presumably have exposed hydrophobic side chains and are pictured as partially or wholly embedded in the lipid bilayer. These proteins are extracted from membranes with difficulty. Indeed, they may require a lipid environment for optimal activity.

Membrane Lipids

A survey of the lipid composition of mitochondrial membranes reveals great differences depending upon the source of mitochondria. McCarty *et al.*, investigated the phospholipid composition of the inner and outer membranes of mung bean (*Phaseolus aureus*) and potato tuber (*Solanum tuberosum*) mitochondria which were prepared in such a way as to exclude contaminating particles. Phospholipids comprised 90% or more of the mitochondrial membrane lipids. The main phospholipids were phosphatidyl choline, phosphatidyl ethanolamine and phosphatidyl glycerol, except in the outer membrane which did not contain significant amounts of phosphatidyl glycerol.

The phospholipid composition is shown in Table 1 together with that of beef heart mitochondria for comparison. Minor amounts of lysophosphatidyl ethanolamine, phosphatidyl inositol, and phosphatidyl glycerol were also found. The fatty acids of the three principal phospholipids of mung bean mitochondria were palmitic, linoleic and linolenic acids. Stearic acid occurred in conjunction with phosphatidyl glycerol. Schwertner and Biale, by contrast, found that phospholipids comprised only about 50% of the total lipids of mitochondria from avocado (*Persea* sp.), cauliflower (*Brassica oleracea*) and potato tubers, and found more phosphatidyl inositol than phosphatidyl glycerol.

Table 1. Phospholipid composition of mitochondrial membranes
% total lipids

Phospholipid	*mung bean inner membrane*	*potato inner membrane*	*potato outer membrane*	*beef heart mito-chondria*
Phosphatidyl choline	33	33	36	38
Phosphatidyl ethanolamine	32	33	64	37
Phosphatidyl glycerol	23	19	none detected	16

The principal fatty acids of the phospholipids of avocado, cauliflower and potato mitochondrial membranes were palmitic, linoleic and linolenic. The fatty acids of the neutral lipids of cauliflower mitochondria were C_{16} and shorter chain acids, while those of potato mitochondria included palmitic, oleic, linoleic and linolenic. The fatty acid composition of the membrane lipids can be highly variable and may reflect the growth conditions of the tissue. Mitochondria from cold grown mung beans (15°C) have a higher amount of unsaturated fatty acids than mitochondria isolated from mung beans grown at 25°C.

Sterols comprise about 2.6% of the total lipids of maize shoot mitochondria. These may be esterified, of which cholesterol and *b*-sitosterol are the principal compounds. Of the unesterified sterols, stigmasterol and *b*-sitosterol are the principal ones. The fatty acids esterified to the sterols include lauric, myristic, palmitic, and linolenic as the most abundant.

Enzymes

Enzymes of the Tricarboxylic Acid Cycle

Mitochondria contain all the enzymes of the tricarboxylic acid cycle. Isolated mitochondria oxidise all of the acids of the cycle, and chromatographic analysis shows that the products are those expected for the reactions of the tricarboxylic acid cycle.

Citrate synthetase

Citrate synthetase (Citrate oxalacetate-lyase (CoA-acetylating)) activity is associated with a particulate fraction of leaf and root tissue from tobacco, bean, and soybean, which sediments at 10,000 g. This fraction consisted largely of mitochondria. Citrate was formed in the presence of acetyl CoA and oxalacetate. Citrate synthetase has been isolated from wheat scutellum mitochondria with minimal contamination by glyoxysomes. The mitochondrial citrate synthetase was released by sonic disruption of the mitochondria and recovered in the supernatent fluid. The purified enzyme had a molecular weight of 96,000 daltons as determined by its elution volume in Sephadex G-100 gel filtration. The K_m for acetyl CoA and for oxalacetate were 4 μM and 34 μM respectively. The enzyme was inhibited competitively with respect to acetyl CoA by ATP. In these respects, the mitochondrial citrate synthetase from wheat scutellum mitochondria is similar to that from mammalian sources.

Pyruvate oxidase

Pyruvate is oxidised by mitochondria with a requirement for catalytic amounts of one of the TCA cycle acids. TCA cycle acids were added to castor bean (*Ricinus communis*) mitochondria at a concentration of 0.001 M . Oxygen consumption ceased after 60 minutes. When pyruvate at a concentration of 0.01 M was then added, oxygen consumption continued at rapid rates and linearly up to three hours. Other cofactors required were NAD, coenzyme A, ATP, and thiamin pyrophosphate.

Isocitrate dehydrogenase

A NAD-specific isocitrate dehydrogenase [L-isocitrate:NAD oxidoreductase (decarboxylating)] has been purified from pea shoot (*Pisum sativum* var Alaska) mitochondria. The enzyme was released from pea shoot mitochondria ruptured by extrusion through a French pressure cell at 3,000 lbs in^{-2}. The enzyme, whose K_m for NAD was 0.22 μM, was activated by Mn^{2+} and Mg^{2+} and to a lesser extent by Zn^{2+}, and inhibited by NADH, with K_i = 0.19 mM. The mitochondrial isocitrate dehydrogenase was specific for NAD which differentiates it from the cytosolic isocitrate dehydrogenase, which requires NADP as the electron acceptor.

Malate dehydrogenase

Malate dehydrogenase (L-malate:NAD oxidoreductase) exists in both mitochondrial and cytosolic forms. Moreover, the mitochondrial malate dehydrogenase may occur as several isozymes. Ting *et al.*, separated two isozymes from young maize (*Zea mays*) mitochondria after sonic disruption. Starch gel electrophoresis revealed a faint fast moving band and a major slow moving band.

Grimwood and McDaniel also found a major slow moving band in polyacrylamide gel electrophoresis with several lighter fast moving bands. Boulter and Laycock attributed the minor bands to complexes of the mitochondrial malate dehydrogenase with other proteins, since re-electrophoresis of the eluted bands always gave a band in the main mitochondrial fraction as well as the original minor band. They determined the molecular weight of the main malate dehydrogenase to be 74,000 daltons.

Plant mitochondria oxidise malate readily, but glutamate must be included in the reaction mixture to remove the accumulated oxalacetate, due to the unfavourable equilibrium of the reaction. The oxidation of malate with endogenous NAD^+ is inhibited by rotenone and antimycin, but is insensitive to these inhibitors when exogenous NAD^+ is supplied.

Malic enzyme

Malate may be oxidised by a NAD^+-dependent malic enzyme [L-malate:NAD oxidoreductase (decarboxylating)] with the formation of pyruvate. Macrae and Moorhouse showed that pyruvate accumulated during malate oxidation by cauliflower bud mitochondria, unless thiamin pyrophosphate was included in the reaction medium. Under the latter conditions, malate was probably oxidised by both malic enzyme and malate dehydrogenase so that in the presence of cofactors for pyruvate oxidation and citrate formation, the latter accumulates.

Malic enzyme is the main pathway for malate oxidation by wheat shoot mitochondria, since oxalacetate did not inhibit malate oxidation, although it did inhibit transiently the oxidation of citrate and pyruvate.

The activities of malic enzyme and malate dehydrogenase differ in mitochondria from various sources. While the relative activity of malate dehydrogenase has in all cases been greater than the activity of malic enzyme, the accumulation of pyruvate *vs* oxalacetate may vary considerably. Pyruvate was accumulated in preference to oxalacetate by a ratio of 27.0 in cauliflower bud mitochondria, while the pyruvate/ oxalacetate ratio for wheat shoot mitochondria was 0.13. When malic enzyme activity is high, mitochondrial NADH levels are raised, and thereby reduce oxalacetate accumulation by product inhibition of malate dehydrogenase.

A strong pH dependence of the activities of the two enzymes was also observed. At pH 6.0 to 7.0, pyruvate accumulates, while at pH values between 7.0 and 8.0 pyruvate accumulation drops and oxalacetate accumulation rises. The pathway may reflect the pH profiles of malic enzyme and malate dehydrogenase. Below pH 7.0, the activity of malic enzyme would maintain a high internal concentration of NADH which would favour the conversion of oxalacetate to malate; above pH 7.0, the decreased activity of malic enzyme and the consequent drop in the NADH levels would favour the oxidation of malate by malate dehydrogenase.

Enzymes of fatty acid oxidation

Fatty acids were oxidised by a particulate preparation from peanut cotyledons. This fraction was identified as the mitochondrial fraction. Cooper and Beevers have separated the particulate fraction from castor bean and have shown that the enzymes of the *b*-oxidation pathway as well as the enzymes of the glyoxylate pathway are associated instead with a heavy particle, the glyoxysome. Mitochondria from castor beans contained less than 5% of the glyoxylate cycle enzymes and virtually none of the *b*-oxidation enzymes.

Enzymes of Fatty Acid Biosynthesis

Isolated mitochondria from avocado mesocarp, flowerlets of cauliflower, and from white potato tubers were capable of incorporating radioactive acetate, acetyl-CoA, malonate, or malonyl-CoA into long-chain fatty acids. Cofactor requirements included coenzyme A, NADPH, ATP, and Mg^{2+} or Mn^{2+}. The principal acids formed were palmitic and stearic acids by avocado mesocarp mitochondria, while mitochondria from cauliflower and white potato tubers synthesized some mono-unsaturated fatty acids as well, i.e., palmitoleic (9-hexadecanoic acid) and, with longer incubation times, oleic (9-octadecanoic) and cis-vaccenic (11-octadecanoic) acids. With the appearance of the C_{18} mono-unsaturated acids, stearic acid is found only in trace amounts, indicating that the unsaturated C_{18} acids were formed from stearic acid.

Enzymes of Phospholipid Biosynthesis

The synthesis of phospholipids proceeds *via* the following reaction:

$$\text{Glycerol} - 3 - \text{phosphate} \xrightarrow[\text{+fatty acids}]{\text{+CoASH, + ATP}} \text{Phosphatidic acid} \quad (1)$$

or

$$\text{Diglyceride + ATP} \rightarrow \text{Phosphatidic acid} \quad \text{(1a)}$$

$$\text{Phosphatidic acid + CTP} \rightarrow \text{CDP - diglyceride} + PP_i \quad (2)$$

Mitochondria isolated from flowerlets of cauliflower contain all of the enzymes necessary for the formation of CDP-diglyceride from glycerol-3-phosphate, when coenzyme A, ATP, CTP and fatty acids are provided. Radioactivity from ^{32}P-ATP was found in phosphatidic acid in peanut cotyledon mitochondria. The incorporation of ^{32} *p* into phosphatidic acid was stimulated by the presence of small amounts of *a* ,*b*-diglyceride, indicating the presence of a mitochondrial diglyceride phosphokinase. ^{14}C-CTP was incorporated into CDP-diglyceride by cauliflower mitochondria. The radioactivity of CDP-diglyceride declines in the presence of *a*-glycerol phosphate or inositol, with the expected formation of phosphatidyl glycerol phosphate or phosphatidyl inositol. Using preparations carefully purified in sucrose density gradients from mung bean hypocotyl mitochondria, Douce *et al.* showed that the CTP:phosphatidic acid cytidyl transferase activity was associated with the inner membrane fraction. Since the activity was not released upon sonication, it was assumed that the transferase was a membrane bound enzyme.

Mitochondrial Electron Transport

Components of the respiratory chain

Reducing equivalents derived from the oxidation of the TCA cycle acids are oxidised in a stepwise manner in the mitochondrial electron transport chain. The electron transport chain, or respiratory chain, is a series of functionally linked election carriers which undergo alternate reduction and oxidation, with molecular oxygen as the terminal election acceptor. Elections are donated by carriers of low redox potential to carriers of high redox potential. It is via these oxidation-reduction reactions that the main oxidative cellular energy transduction occurs, either through the formation of intermediate states which can be coupled to cellular work, (e.g., ion transport) or through the phosphorylation of ADP to form ATP, which can then mediate cellular endergonic reactions.

The main components of the respiratory chain have been identified, both by characteristic reaction toward inhibitors and by spectral analysis. In its basic form, the respiratory chain in plant mitochondria is very similar to that of mitochondria from fungal or animal sources. The chain, (Fig.1) can be functionally separated into (a) NADH: coenzyme Q oxidoreductase; (b) succinate: coenzyme Q oxidoreductase; (c) reduced coenzyme Q: cytochrome *c* oxidoreductase; and (d) cytochrome oxidase.

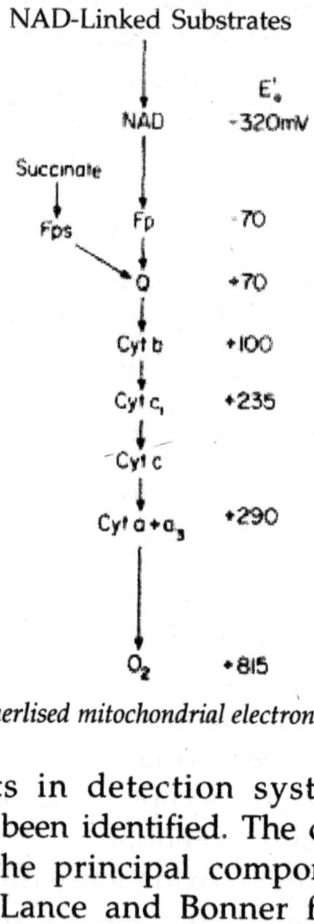

Figure 1. The generlised mitochondrial electron transport chain.

With refinements in detection systems, additional components have been identified. The concentration and molar ratios of the principal components have been determinded by Lance and Bonner for mitochondria isolated from a number of sources and are given in Table 2. The concentrations of the cytochromes are quite similar to those of animal mitochondria.

Nicotinamide Adenine Dinucleotide

Malate dehydrogenase and isocitrate dehydrogenase are NAD^+ linked enzymes. The oxidation of substrates is accompanied by the reduction of endogenous NAD. This is shown by the strong fluorescence of reduced NAD. Carefully isolated mitochondria contain sufficient endogenous NAD to oxidise malate or isocitrate. The oxidation of malate or isocitrate by endogenous NAD is inhibited by rotenone or amytal. Malate oxidation is stimulated by the addition of NAD, but the oxidation then becomes insensitive to rotenone inhibition, just as the oxidation of exogenous NADH is insensitive to rotenone or amytal suggesting more than one pathway of NADH oxidation. Douce *et al.*, and Day and Wiskich delineated three mitochondrial NADH dehydrogenases, one located on the outer membrane, a second located on the outer surface of the inner membrane, and a third on the inner surface of the inner membrane. Each dehydrogenase has a characteristic response to inhibitors. The outer membrane dehydrogenase is characterised by an antimycin insensitive NADH: cytochrome c reductase with added cytochrome *c*. In intact mitochondria, the NADH dehydrogenase of the outer surface of the inner membrane is coupled to cytochrome oxidase and goes through the antimycin sensitive site, but by-passes the rotenone sensitive site.

This dehydrogenase shows NADH: cytochrome *c* reductase activity at low osmolarity only, due to the impermeability of the intact outer membrane to added cytochrome *c*. These first two dehydrogenases oxidise exogenous NADH. The NADH dehydrogenase of the inner surface of the inner membrane oxidises the NADH linked to malate and isocitrate dehydrogenases. Electrons must go through the rotenone and the antimycin sensitive sites to cytochrome oxidase. In the intact mitochondrion, these various pathways interact, as demonstrated by the relief of antimycin or rotenone inhibition of malate: cytochrome *c* reductase activity by added NAD^+. Further evidence for the delineation of the NADH dehydrogenases is obtained through the P/O or ADP/O ratios of tightly-coupled mitochondria. Oxidation of malate or isocitrate gives ratios approaching 3.0, while the oxidation of NADH gives ratios approaching 2.0. The by-pass of the rotenone sensitive portion of the respiratory chain results in by-passing one of the phosphorylation sites as well.

Flavoproteins of NADH dehydrogenase

Storey distinguished a flavoprotein, F_P M , which was rapidly reduced upon addition of malate. The reduction of this flavoprotein was inhibited by amytal. He tentatively assigned a midpoint potential, $E_{m7.2}$ = –70mV for this flavoprotein. F_{PM} is the flavoprotein involved in the first energy conservation site and hence is most likely the flavoprotein involved in the NADH dehydrogenase located in the inner surface of the inner membrane.

F_{Pha} , a high potential non-fluorescent flavoprotein is rapidly reduced upon the addition of exogenous NADH in mung bean mitochondria. This reduction is insensitive to amytal. Its midpoint potential is approximately +110mV. Flavoprotein F_{Pha} could be the flavoprotein associated with the dehydrogenases which oxidise exogenous NADH, but it is not possible to determine if it is the flavoprotein of the outer membrane dehydrogenase or the inner membrane dehydrogenase from the information available.

Flavoprotein of succinate dehydrogenase

Isolated mitochondria from a number of plant sources oxidise succinate readily, with oxygen consumption rates of about 450 nM O_2 min^{-1} mg^{-1} protein. Activation by ATP is required to obtain maximal rates of respiration with succinate as substrate, as well as for rapid response to addition of ADP. The activation by ATP is often attributed to the removal of inhibitory amounts of bound oxalacetate but the mechanism of the activation is not clear. The ATP effect is not due to phosphorylation mechanisms since neither oligomycin nor dinitrophenol affect the ATP activation. Hiatt reported the partial purification of succinate dehydrogenase from mitochondria from bean roots and tobacco (*Nicotiana tabacum*) leaves. The K_m of the enzyme for succinate was 1 mM.

Malonate inhibits competitively, with $K_i = 0.24$ mM. The apparent Michaelis constant of isolated mitochondria for succinate, in the presence of ADP and phosphate was 0.4 mM. Singer *et al.*, studied the succinate dehydrogenase of submitochondrial particles prepared by sonication of isolated mung bean and cauliflower mitochondria. The enzyme in submitochondrial particles was activated by a number of agents including substrate or Br^-. With Br^- activation, oxalacetate was removed, although it cannot be assumed that the oxalacetate was uniquely associated with the succinate dehydrogenase of the particles. Other activators included CoQ_{10} , NADH, NAD-linked substrate (i.e., malate plus pyruvate), and ADP. Succinate reducible flavoprotein was not detected spectrally, but a flavoprotein associated with succinate dehydrogenase was determined chemically. Singer *et al.*, found that succinate dehydrogenase contained covalently bound flavin as a histidyl-*a*-FAD. The flavin content was approximately 0.2 nM per mg protein. The molar ratio of flavin to enzyme was not determined.

Ubiquinone

Studies on the role of ubiquinone in the plant mitochondrial electron transport chain have not been extensive. Beyer *et al.*, extracted ubiquinone from mung bean submitochondrial particles. A single ubiquinone was found which co-chromatographed with ubiquinone-10. The spectrum of the extracted ubiquinone-10 has an absorption peak at 275 nm in the oxidised form and at 290 in the reduced form. The ubiquinone was reduced by succinate and NADH; at an aerobic steady state, 38% of the ubiquinone was reduced by succinate, while 56% was reduced by NADH. At anaerobiosis induced by succinate or NADH, 88% and 84% respectively were reduced.

Sodium hydrosulphite gave additional reduction. The quinones were virtually 100% oxidised in aerobic suspension in the absence of substrates. Ubiquinone is generally acknowledged as part of the mitochondrial respiratory chain and is placed at the juncture of succinate dehydrogenase and NADH dehydrogenase. This placement is based on the considerable work with animal mitochondria. Storey and Bahr, basing their conclusions upon measurements of the half times of reduced to oxidised transitions, and the times for 50% reduced to the fully reduced state, suggested that ubiquinone is in the main respiratory chain of mung bean mitochondria. Ubiquinone is the link between the dehydrogenases and the cytochromes, as usually regarded in animal mitochondria, but Storey and Bahr in addition placed F_{Pha}, the high potential flavoprotein, between ubiquinone and the *b*-cytochromes. F_{Pha} has no counter-part in the animal mitochondrial respiratory chain.

Cytochrome b

The cytochromes are hemo-proteins. Three classes of cytochromes, distinguished by their spectral properties, as well as by the nature of their prosthetic groups are found in mitochondria. The *a*-cytochrome contains as its prosthetic group, heme *a*, while the *b*- and *c*-cytochromes contain a heme closely related to protoporphyrin IX. In the *c*-cytochromes, the heme is covalently linked to the protein *via* sulphur atoms in a thio-ether linkage. In the reduced state, the cytochromes exhibit strong absorption bands in the visible region of the spectrum which have been useful in their identification and in the analysis of their function.

In addition, both the oxidised and reduced forms absorb strongly in the region around 400 nm, which is a characteristic of all heme compounds.

The *b*-cytochromes are best resolved when their spectra are determined at low temperatures, e.g., 77°K. Three *b*-cytochromes have been identified and two others are suggested. Their spectral properties are summarised in Table 2.

Table 3. The b-cytochromes of plant mitochondria.

Cytochrome	*a-peak,* *25°C nm*	*a-peak,* *77°K nm*	E_m *mV*
b–556 556	553–554	+75 to + 100	
		553	+ 75
b–560 560	557	+ 40 to + 80	
		557	+42
b–558 557–558	553–555	–70 to –105	
b–566 566	561–563	–75	
		562	–77
$S_2O_4^{2-}$ reducible	557–561	554, 560	–100

Considerable variability exists in the nomenclature of the *b*-cytochromes. In conformity to the International Union of Biochemistry, the *b*-cytochromes are designated according to the *a*-peak of their reduced spectrum at room temperature (25°C). It should be noted that there is a blue shift of about 3 nm in the spectrum at 77°K relative to the spectrum at room temperature. The use of the *a*-absorption peak is further complicated by the fact that some authors use the absorption maximum at 77°K to designate the various *b*-cytochromes. In older nomenclature, mammalian cytochrome *b*–562, as orginally described by Keilin, was designated cytochrome *b*. As other *b*-cytochromes were discovered with an *a*-absorption peak significantly different from 562 nm, these were designated with subscripts. More recently, cytochrome *b*–566 was thought to be directly involved in energy transduction and was designated cytochrome b_T, a transducing *b*-cytochrome, to differentiate it from cytochrome *b*–562, or b_K.

The multiplicity of *b*-cytochromes is due to different *b*-cytochromes in mitochondria rather than a splitting of the absorption bands at low temperature, since the peak heights do not change in synchrony in the presence of reducing agents, inhibitors or uncouplers. The *b*-cytochromes are placed in the respiratory chain according to the following sequence:

$$\text{Cyt b-556} \rightarrow \text{cyt b-560} \rightarrow \text{cyt c-55t}$$

Cytochrome *b*–560 (557) was placed on the oxygen side of cytochrome *b*–556 (553) as a result of the determination of the rates of oxidation of the 556 and 560 components by an oxygen pulse of an anaerobic suspension of mitochondria. The 560 component was oxidised with a half-time of oxidation of 6 to 8 msec while the 556 component was oxidised with a half-time of 150 to 200 msec. The reduction by succinate of these two components in anaerobiosis showed, however, that *b*–560 was reduced more slowly than *b*–556 which was contrary to the expected rates in view of the rates of oxidation. The slow reduction was ascribed to the more negative redox potential of *b*–560. The midpoint potentials of *b*–560 and *b* –556 would predict that *b*–560 would be on the substrate side of *b* –556. Further resolution of the sequence of the *b*-cytochromes is necessary.

Cytochrome *b*–566 was thought to be analogous to the *b*–566 (b_T) of mammalian mitochondria. Cytochrome *b*–566 from animal mitochondria was found to undergo a midpoint potential shift as well as an enhanced reduction in anaerobic suspension when the respiratory chain was energised. This was interpreted as the formation of a high energy intermediate of phosphorylation directly involving cytochrome *b*–566. In plant mitochondria, the midpoint potential shift of *b*–566 was not observed. Although enhanced reduction of *b*–566 by ATP or by energisation of the respiratory chain could be demonstrated in plant mitochondria, it could be explicable by reverse electron flow through the *b*-cytochromes. Thus the status of a transducing *b*-cytochrome in plant mitochondria is in question. In fact, cytochrome *b* –566 was excluded from the main sequence of the respiratory chain, since it remains oxidised in anaerobic suspensions while other *b*-components, pyridine nucleotides and fluorescent flavoproteins are reduced. There was a lack of equilibration between the low redox potential carriers with cytochrome *b*–566. The function of cytochrome *b* –566 is left uncertain.

Cytochrome c

Two *c*-type cytochromes have been detected in plant mitochondria. They have the same relationship as cytochrome *c* and c_1 in animal mitochondria. The room temperature spectrum shows a large peak at 550 mn which shifts to 547 nm at liquid nitrogen temperature (77°K). As with cytochrome c in animal mitochondria, the cytochrome c (cyt c –547[*]) is easily extracted by salt solutions. A second component with a low temperature absorption peak at 549 nm remains after extensive washing with phosphate buffer. The 547 nm absorbing component is recovered in the phosphate buffer extract while the 549 nm absorbing component remains in the pellet. Both are reducible by ascorbate, which differentiates them from the b-cytochromes. The spectral properties of the 549 absorbing c component and its strong binding to mitochondrial membranes relate this c component to cytochrome c_1 of animal and yeast mitochondria. The midpoint potentials of the two c-cytochromes of mung bean mitochondria have been determined to be +235 mV in both cases. The half-time of oxidation of cytochrome c–547 and c –549 are 3.0 and 3.1

msec respectively when KCN treated anaerobic mitochondria were pulsed with 14 μM O_2 . The electron transfer sequence of the c-cytochromes was given as cyt c–549 to cyt c–547.

Cytochrome *c* (cyt *c*–547) is essentially identical to cytochrorne *c* from all eukaryotic sources. Cytochrome *c* from one source will react with the reductase and oxidase from quite distantly related sources. The amino acid sequence of cytochrome *c* from a large number of sources is shown to have a high degree of homology. This homology is all the more striking when the tertiary structure is considered. For example, the amino acids about the heme show a high degree of conservatism among the cytochrome *c* proteins examined. Those amino acid residues important to the structure and function of the protein have suffered few substitutions in the course of evolutionary history.

Cytochrome oxidase

The cytochrome oxidase of plant mitochondria contains cytochromes *a* and a_3, as does that from animal mitochondria. These are two spectroscopically differentiated components, although two separate chemically different entities have not been isolated. The optical properties are well differentiated in the presence of cyanide or azide, which binds to cytochrome a_3. The *a*-band of cytochrome oxidase at room temperature is located at 602 nm; at 77°K, there is a blue shift to 598 nm. The reduced spectrum of cytochrome a is revealed in the difference spectrum of an azide treated aerobic suspension minus an aerobic suspension. All components are oxidised except for cytochrome *a*, which shows a symmetrical reduced *a*-band at 598 nm. The a_3 spectrum is shown in a difference spectrum of an anaerobic plus azide suspension, minus an aerobic plus azide suspension. The reduced *a*-peak of cytochrome *a* cancels and the reduced *a*-peak of cytochrome

a_3 is shown with its maximulm at 603 nm. The midpoint potentials of cytochromes *a* and a_3 are +190 and +380 mV respectively. Half-times of oxidation in oxygen pulsed anaerobic suspension are 2.0 msec and 0.8 msec respectively for cytochromes *a* and a_3.

Non-Heme iron proteins

Few investigations have been carried out on the occurrence and nature of non-heme iron proteins in higher plant mitochondria. Electron paramagnetic resonance (epr) signals characteristic of iron sulphur centres were observed by Cammack and Palmer in mitochondria from *Helianthus tuberosus* and *Arum maculatum* with components at g = 2.02 and 1.93, 2.05 and 1.92, and at 2.10 and 1.87. Schonbaum *et al.*, obtained an epr signal in NADH-reduced skunk cabbage (*Symplocarpus foetidus*) mitochondria for a component at g = 1.94 and some complex components near g = 2.00. These iron sulphur centres no doubt participate in electron transport, since they undergo oxidation and reduction. At present, their precise functions are not known. They may be analogous to the iron sulphur centres identified in the NADH dehydrogenase segment of yeast and pigeon heart submitochondrial particles, which are believed to be closely involved in energy coupling.

Cyanide Resistant Respiration

An unusual characteristic of respiration in plant mitochondria is a partial insensitivity to cyanide inhibition. Partial insensitivity is exhibited to inhibition by azide, antimycin and 2-heptyl hydroxy-quinolin-N-oxide, all of which are potent inhibitors of oxygen uptake in animal mitochondria. This cyanide insensitivity may be almost 100% as in the spadix mitochondria from some aroid species, notably of *Arum maculatum* and *Symplocarpus foetidus*, partial as in mung bean mitochondria, or completely lacking as in the mitochondria from fresh, dormant white potato tubers. In the latter, a cyanide insensitivity, and indeed a cyanide stimulation of respiration, may be induced upon vigorously aerating slices of potato tuber tissue in water for 24 hours. Mitochondria isolated from such aged potato tuber slices are much less inhibited by antimycin or cyanide.

Because of the insensitivity to cyanide at concentrations which completely inhibit the respiration of animal mitochondria, it is unlikely that the cyanide insensitive respiration is due to an incomplete inhibition of cytochrome oxidase. This 'excess' oxidase hypothesis has been considered by various workers and received strong support from the finding that cytochromes *a* and *c* were incompletely reduced in the presence of cyanide. The insensitivity to antimycin, which inhibits electron transport between the cytochrome *b* to cytochrome *c* region favours the argument that the cytochrome system is by-passed entirely and that there exists an alternate oxidase in mitochondria showing cyanide insensitivity, although mitochondria of *A. maculatum* and *S. foetidus* have the conventional cytochrome complement.

By and large, cytochrome *c* and cytochrome oxidase are reduced in the presence of cyanide. The incomplete reduction of cytochromes found by Chance and Hackett could be attributed to (a) the fact that the cytochromes are not always reduced by substrates relative to the reduction by dithionite; (b) possible spectral interference in the measurement of cytochrome *c* due to an oxidised *b*-cytochrome; or (c) in coupled mitochondria, significant reverse electron transport may cause the carriers on the oxygen side of the respiratory chain coupling site to become partially oxidised. Chance and Hackett and Bendall and Hill reported, however, that a *b* –type cytochrome becomes oxidised in the presence of cyanide and oxygen. Bendall and Hill called their *b*-component from *A. maculatum* cytochrome b_7 (a_{max} 560 nm), while Chance and Hackett identified an oxidisable *b*-component with an a_{max} at 558.5 nm.

It was hypothesized that a *b*-cytochrome functioned as the shunt to the alternate oxidase. Such a role for a *b*-cytochrome is favoured, since Bahr and Bonner reported that the known flavoproteins and ubiquinone had equal access to oxygen by either pathway, based on observations of their oxidation in the presence or absence of cyanide, and Storey and Bahr found no identifiable carriers among flavoprotein, ubiquinone, the known *b*-cytochromes or *c*-cytochromes which could mediate electron transfer to the alternate oxidase. Cytochrome b_7 was identified with cytochrome *b*–557 (77°K), but since the oxidation rate of cytochrome *b*–557 was unaffected by m-chloro-benzhydroxamic acid, an inhibitor of the alternate oxidase it was felt that cytochrome *b*–557 does not play a part in the alternate pathway, and should not be equated with Bendall and Hill's cytochrome b_7. Bendall and Bonner showed that thiocyanate and other metal binding agents also inhibit the alternate pathway, suggesting a role for non-heme iron proteins or other metalloproteins. Efforts to identify non-heme iron proteins which may be the alternate oxidase have not been successful.

Strong epr signals characteristic of iron-sulphur proteins in mitochondria of aged Jerusalem artichoke (*H. tuberosus*) and in *A. maculatum* mitochondria were found. However, these are most likely the iron sulphur proteins of NADH: ubiquinone reductase since they were not reducible by succinate, and were, moreover, unaffected by hydroxamic acids. Schonbaum *et al.,* found that the epr signals in NADH reduced skunk cabbage submitochondrial particles were enhanced by treatment with m-iodobenzhydroxamic acid. In addition to a signal at g = 1.94, which is characteristic of NADH dehydrogenase, a set of complex signals near g = 2.0 was detected at 77°K. It was thought that the g = 2.0 signal may originate from the alternate pathway.

The possibility that the alternate oxidase may involve a flavoprotein has been explored. Flavoprotein oxidases which reduce oxygen with the formation of hydrogen peroxide are known, but such oxidases have not been identified for cyanide resistant mitochondria.

The function of cyanide resistant respiration is unclear. In the spadix tissue of maturing flowers of *Arum* and *Symplocarpus*, it may serve the function of thermogenesis. A rise in temperature of the spadix tissue ten degrees above the ambient has been recorded. This thermogenesis may aid these plants in pollination, since flowering occurs in early spring.

Energy Linked Reactions of Mitochondria

Oxidative phosphorylation

Coupling sites

Oxidative phosphorylation, the process whereby the reaction

$$ADP + P_i \rightarrow ATP + H_2O$$

is coupled to the oxidation of reduced electron carriers of the respiratory chain occurs in plant mitochondria in the same manner as does the reaction in animal mitochondria. The sites of phosphorylation are identical. These are the sites I in the NADH: ubiquinone reductase segment, II in the cytochrome *b*: cytochrome *c* reductase segment and III in the cytochrome oxidase segment of the respiratory chain. It has not been possible to identify precisely those components of the respiratory chain which are the energy transducers in each of the three sites, although these have been approximated from the change in redox potentials (*D* E_0') between adjacent carriers, by application of the crossover theorem, and from shifts in the midpoint potential of certain electron carriers upon the addition of

ATP to uncoupled anaerobic suspensions of mitochondria. The *D* E_0' indicates the thermodynamic feasibility of coupling between two adjacent components.

Thus, coupling sites were thought to be located between endogenous NAD and the flavoprotein of NADH: ubiquinone reductase; between cytochrome *b* and cytochrome *c* of the cytochrome *b* :cytochrome *c* reductase; and between cytochrome *a* and oxygen of cytochrome oxidase. The application of the crossover theorem by and large confirms the general location based on thermodynamic grounds. According to the crossover theorem, electron transport through the coupling site is the rate limiting step in tightly coupled mitochondria. The carrier on the substrate side of the coupling site should become reduced, while the carrier on the oxygen side would become oxidised. Upon the addition of the phosphate acceptor, ADP, there would be observed a rapid transient oxidation of the carrier on the sub-strate side, and a reduction of the carrier on the oxygen side of the coupling site, concomitant with the release of controlled respiration in the state 4-state 3 transition (Table 3).

Table 3. Definition and values of steady states in mitochondria

State	*(O_2)*	*ADP levels*	*Substrate*	*Respiratory rates*	*Rate-limiting substance*
1	> 0	low	low	slow	ADP
2	> 0	high	~0	slow	substrate
3	> 0	high	high	fast	respiratory chain
4	> 0	low	high	slow	ADP
5	0	high	high	0	oxygen

Wilson and Dutton reported that the midpoint potential (E_m) of cytochrome a_3 becomes more negative upon energisation of an anaerobic suspension of rat liver mitochondria by ATP. Similarly they report that the E_m of one of the *b* cytochromes (a_{max} 564 nm) increases upon energisation by ATP. These shifts of the midpoint potentials were attributed to changes in ligand interaction energy of the heme iron upon energisation, and that the iron atoms of these cytochromes were directly involved in energy coupling.

Such a change in the midpoint potential was postulated by Wang. In his model, the phosphoimidazole group becomes a much weaker coordinating ligand for the Fe(II) than imidazole, and hence should lower the midpoint reduction potential of the corresponding electron carrier. Similar experiments in the iron-sulphur centres of the site I region of the respiratory chain showed that of the five iron sulphur centres identified, ATP affected the reduction potential of only one of these centres, designated as centre I. Addition of ATP caused a partial oxidation of centre I when the reduction potential was poised at a value where the iron was almost completely reduced.

These observations suggested that the reduction potential of centre I was dependent upon the phosphate potential, and that the addition of ATP caused a lowering of the midpoint potential of the centre I iron sulphur protein. The significance of the changes in the midpoint potential was questioned, at least for the site II region, since none of the *b*-cytochromes showed an ATP-induced potential shift when investigated in plant mitochondria. One must conclude that there is a fundamental difference in the function of the *b*-cytochromes in plant and animal mitochondria or that the changes in the midpoint potential do not reflect the formation of an energy transducing species. Lambowitz *et al.*, took the latter view and attributed the midpoint potential changes to a reverse electron flow with cytochrome *b* –566 and cytochrome a_3 equilibrating with the redox mediators by way of cytochrome *c*, while the iron sulphur centre of site I equilibrated through the nicotinamide adenine nucleotide pool.

ADP:O Ratios

The general location of the three coupling sites means that for tightly coupled mitochondria, predictable ratios of the moles of ADP phosphorylated to the gram-atoms of oxygen consumed (ADP/O or P/O) may be obtained for each substrate. Thus succinate should give a ratio of 2.0, NAD-linked substrates a ratio of 3.0 and *a*-ketoglutarate a ratio of 4.0 (one phosphorylation *via* succinyl-CoA, three phosphorylations *via* the respiratory chain NADH). These ratios are largely confirmed in plant mitochondria. Early attempts gave results much below the predicted ratios, but with refinements in the procedure for the isolation of plant mitochondria, the inclusion of required cofactors, and the exclusion of competing reactions, predicted ratios have been approached, as shown in Table 4.

Table 4. Oxidative phosphorylation by isolated mitochondria.

Source	*Substrate*	*P:O*	*Respiratory control index*
Phaseolus aureus	succinate	1.68	3.2
	malate	2.56	5.4
	NADH	1.55	4.5
Solanum tuberosum	succinate	1.64	3.4
	malate	2.59	4.2
	NADH	1.48	2.4

Exogenous NADH does not interact with respiratory chain NAD, but with flavoproteins of the external dehydrogenase. These ultimately enter the respiratory chain at the level of cytochrome *b* and hence electrons go through two coupling sites only.

Energy coupling in cyanide resistant respiration

Mitochondria of *Symplocarpus* are fully capable of energy conservation in the uninhibited state. Hackett and Haas found P/O ratios greater than 3 for *a*-ketoglutarate oxidation by skunk cabbage mitochondria. In the presence of cyanide, the energy coupling sites associated with the cytochromes are by-passed, and energy coupling involves only the NADH:ubiquinone reductase portion of the respiratory chain. Storey and Bahr obtained ADP/O ratios of 1.3 with succinate as substrate in skunk cabbage mitochondria in the absence of cynide; the ratio was zero in the presence of cyanide. With malate as the substrate, the ADP/O ratios were 1.9 and 0.7 in the absence and presence of 0.3 mM KCN respectively, showing that coupling sites II and III were effectively by-passed while coupling site I, which could be activated by malate but not by succinate, was still functioning in the presence of cyanide.

Wilson, however, has proposed that energy coupling occurs in the alternate pathway as well. He found that cyanide treated mitochondria of *A. maculatum* and *Acer pseudoplatanus* produced ATP, with an ATP/O ratio of 0.5 at high O_2 concentration with succinate as the substrate. Submitochondrial particles of *A. maculatum* which lacked malate dehydrogenase activity oxidised succinate with P/O ratios of 0.2 to 0.6. Since malate dehydrogenase-depleted particles were still capable of phosphorylation, the possibility that the phosphorylation was due to the oxidation of a product of succinate oxidation, i.e., malate, could be eliminated, therefore suggesting that a phosphorylation site exists in the alternate pathway which functions only at high C_2 concentration.

Cyanide inhibited mitochondria from spadices of *A. maculatum* and from mung bean hypocotyls still retain an energy related function through the cytochromes. Bonner and Bebdall showed that a substrate linked reverse electron flow is active in spadix mitochondria. Cytochrome *c* and cytochrome oxidase of cyanide treated mitochondria were reduced by ascorbate plus, *N,N,N',N'*-tetramethyl-*p*-phenylenediamine (TMPD), an electron donor to cytochrome *c*. Subsequent addition of succinate caused a reoxidation of cytochrome *c* and cytochrome *a* with a partial reduction of cytochrome *b*. The reoxidation of cytochromes *c* and *a* was prevented by uncoupling concentrations of carbonylcyanide-phenyl-hydrazone or 2,4-dinitrophenol. Wilson and Moore also showed that cyanide inhibited mung bean mitochondria could carry out oxidation of ascorbate plus TMPD. The oxidation was thought to involve a reverse electron flow through coupling site II and energized by ATP or the high energy intermediate.

REFERENCES

Bonner W.D., Jr., "Mitochondria and plant respiration", In *Phytochemistry*, vol. 3 (ed. L.P. Miller). Van Nostrand Reinhold, 1973.

Goddard D.R. & Bonner W.D., Jr., "Cellular respiration", In *Plant Physiology*, vol. 1A (ed. F.C. Steward). Academic Press, 1960.

Lehninger A.L., *The Mitochondrion: Molecular Basis of Structure and Function*. W.A. Benjamin, Inc., 1964.

Munn E.A., *The Structure of Mitochondria*, Academic Press, 1974.

Öpik H., "Mitochondria", In *Dynamic Aspects of Plant Ultrastructure* (ed. A.W. Robards), McGraw-Hill, 1974.

Sato S. (ed.), "Mitochondria", In *Selected Papers in Biochemistry*, vol. 10. University Park Press, 1972.

Slater E.C., Zaniuga Z. & Wojtczak L. (eds.), *Biochemistry of Mitochondria*, Academic Press, 1967.

8

Microbodies and Microtubules

MICROBODIES

Microbodies were discovered relatively recently with the advent of the electron microscope. They were first recognised in the early 1950's as small spherical bodies in electron micrographs of mammalian kidney and liver tissue and similar organelles were reported in plant tissues in the early 1960's. Later in that decade they were isolated from plant tissues and from studies of their biochemical functions were recognised as being of major importance in plant cell metabolism. It is perhaps interesting to point out that the physiological significance to plant cell metabolism of major organelles, such as mitochondria and chloroplasts, was recognised long before their component biochemical reactions had been studied in detail, while the discovery and isolation of microbodies allowed some already well known metabolic pathways to be ascribed to a specific organelle.

Two types of microbody with identical structures but with distinct physiological functions have been isolated from plant tissues. They have been called *peroxisomes* and *glyoxysomes* to distinguish their separate functions in cell metabolism. Peroxisomes occur in the leaves of higher plants and are closely associated with chloroplasts. They are the sites for the oxidation of glycollic acid, a product of carbon dioxide fixation. The oxidation of this compound, results in a release of carbon dioxide and oxygen uptake which is called photorespiration. Glyoxysomes on the other hand, occur abundantly during the germination of those seeds which store fats as a reserve material, and contain the enzymes necessary for the breakdown of fatty acids to acetyl-CoA and the synthesis of succinate from acetyl-CoA.

All microbodies appear to contain flavin-dependent oxidases and catalase. The oxidation of a substrate by a flavin-linked oxidase is accompanied by the uptake of oxygen and the production of hydrogen peroxide, which is broken down by catalase to oxygen and water.

$$\begin{pmatrix}\text{Reduced substrate} \\ \text{Oxidized substrate}\end{pmatrix}\begin{pmatrix}\text{FAD} \\ \text{FADH}_2\end{pmatrix}\begin{pmatrix}H_2O_2 \rightarrow H_2O + 1/2O_2 \\ O_2\end{pmatrix}$$

Microbodies, therefore, contribute to the uptake of molecular oxygen by the cell, but unlike mitochondria they do not contain any electron transport system which would be needed for the recovery of energy as ATP.

Structure and Occurrence

Microbodies are usually spherical, but can be ellipsoidal or dumbell shaped, and range from 0.2 to 1.5 μm in cross sectional diameter. Because of the low contrast between the matrix of the microbody and the cytosol they are not detectable by light microscopy but in electron micrographs they can be seen to have a single limiting membrane enclosing a granular matrix of moderate electron density. The core of the microbody commonly contains fibrillar inclusions or a single large crystalline inclusion which may be formed by a reorganisation of the fibrillar material. These crystalline bodies have been shown to have catalase activity. No ribosomes or any form of nucleic acids have been detected in the microbody and hence they are not considered to be capable of self-replication. Since they are found in many tissues in association with the endoplasmic reticulum they are thought to be formed by this structure. Microbodies are therefore structurally very simple and cannot be confused in electron micrographs with any organelle except perhaps lysosomes from which they can be distinguished by cytochemical techniques.

The presence of catalase is a distinguishing feature of microbodies and this can be detected cytochemically using the dye 3,3'-diaminobenzidine (DAB). In fixed sections, catalase remains active and in the presence of hydrogen peroxide the dye is oxidised to give an osmiophyllic electron-dense product in the core of the microbody.

Microbodies have been seen in electron micrographs of many plant tissues. They have been found in the leaves of angiosperms, gymnosperms and bryophytes where they are numerous and are invariably located near, or appressed to, chloroplasts. Microbodies have also been found in yeast and hyphal fungi and in species representing a wide range of algal phyla. In the fat-storing seeds of higher plants they are numerous at certain stages of germination and are found in the cells in close association with *spherosomes* which are large lipid storage bodies. In the roots of higher plants a catalase-containing microbody has been found in association with the endoplasmic reticulum but since it does not contain either glycollate oxidase or enzymes of the glyoxyllate cycle, its function in the cell is not clear.

Isolation

Microbodies appear to be very fragile, since homogenisation of plant tissue damages them to such an extent that their constituent enzymes appear in soluble fractions after removal of chloroplasts and mitochondria by centrifugation. More gentle techniques of tissue breakage however allow microbodies to be isolated at a yield of usually about 10% of the organelles present in the whole tissue as judged by the solubilisation of enzymes presumed to be present in the microbody. The method of breakage depends upon the tissue. Leaves are ground for a brief time with a mortar and pestle at 0–4°C in a buffer containing an osmoticum such as sucrose at a concentration of 0.4 to 0.8 M . Other tissues may be chopped or finely sliced with razor blades in a similar medium. The resulting homogenate is squeezed through cheesecloth and the brei centrifuged at a low speed to remove cell debris. The supernatant is then centrifuged at 6,000 to 10,000 g to obtain a pellet containing broken chloroplasts, microbodies and some mitochondria.

Microbodies are separated from the other organelles by layering this resuspended pellet fraction on a discontinuous or continuous density gradient of sucrose ranging in concentration from 1.3 to 2.5 M, and centrifuging in an ultracentrifuge for 3–4 hr at 40,000 g. Microbodies increase in density during this process by a loss of water and ultimately form a band in the gradient at a density of 1.24 to 1.26 g cm^{-3} . This density is higher than that of other organelles and they are clearly separated from mitochondria which sediment at a desntiy of 1.16 to 1.19 g cm^{-3}. On the separation of the gradient into fractions the microbodies are detected by assaying for specific 'marker' enzymes. Typical marker enzymes for peroxisomes are catalase, glycollate oxidase and hydroxypyruvate reductase, while those for glyoxysomes are catalse, ma late synthetase and isocitrate lyase. Mtitochondria are characterised by the presence of cytochrome *c* or succinic dehydrogenase, and chloroplasts by their chlorophyll pigments.

Much of the activity of the microbody marker enzymes is found in the soluble non-particulate fraction of gradients, indicating that the yield of micro-bodies is low. Despite these low yields, sufficient amounts of intact microbodies have been isolated from several plant tissues to be able to determine their metabolic functions. This has resulted in the distinguishing of at least two types of microbody, peroxisomes and glyoxysomes, with distinctly different enzyme complements and physiological functions.

Glyoxysomes

There are many plant species in which lipid is the main storage material in the cotyledons or endosperm of the seed. During the first few days of seed germination there is a dramatic decrease in the lipid content of the seed and sugars, principally sucrose are formed. These sugars are subsequently translocated to the growing embryo or embryonic axis. This lipid to carbohydrate conversion has been correlated with an increase in the activity of the glyoxysomal enzymes, malate synthetase and isocitrate lyase, and as germination proceeds and the lipid reserves are depleted, the number of lipid storage bodies or spherosomes decrease and there is a drop in the activities of malate synthetase and isocitrate lyase. The elucidation of the metabolic processes involved in this process of *gluconeogenesis,* ultimately led to the isolation of particles in which were localised the crucial enzymes of this pathway. These particles were found to be morphologically similar to animal peroxisomes and were called glyoxysomes.

The breakdown of lipids is initiated by their hydrolysis to fatty acids. Triglycerides are hydrolised to glycerol and fatty acids by the enzyme lipase while phospholipids are hydrolised by phospholipase. The resultant long-chain fatty acids are subsequently degraded by the successive removal of 2-carbon fragments in the process of *b* - oxidation.

Glyoxyllate Cycle

The acetyl-CoA derived from fatty acid breakdown in germinating seeds could be consumed by the TCA cycle, as indicated by the experiments of Stumpf and Barber. However this would not result in the net formation of a glucose precursor since for each molecule of acetyl-CoA consumed two molecules of carbon dioxide would be produced. Furthermore, it was known at the time that little of the fatty acid was oxidised to CO_2 but instead contributed to a net synthesis of sugars.

The problem of how acetyl-CoA was converted to a glucose precursor was solved by the discovery of the glyoxyllate cycle, by Kornberg and Krebs in 1957. This cycle represents a modification of the tricarboxylic acid cycle in which two molecules of acetyl-CoA are consumed and a molecule of succinic acid is formed. Five enzymes are involved in this process three of which, citrate synthetase, aconitase, and malic dehydrogenase are components of the TCA cycle. The remaining two enzymes, the key enzymes of the glyoxyllate cycle, are isocitric lyase (isocitratase) and malate synthetase. The first reaction of the glyoxyllate cycle, catalysed by citrate synthetase, is the condensation of oxaloacetate and acetyl-CoA to form citrate, which is then converted to isocitrate by the action of aconitase. The next reaction, unique to this cycle, is the cleavage of isocitrate into succinate and glyoxyllate catalysed by isocitrate lyase. One of the products of this reaction, glyoxyllate, is then condensed with a second molecule of acetyl-CoA under the catalytic action of malate synthetase, to produce one molecule of malate. Malate is then oxidised by malate dehydrogenase to oxaloacetate with the concomitant reduction of NAD^+. The overall equation for the cycle is therefore:

$$2 \text{ acetyl-CoA} + NAD^+ \rightarrow \text{succinate} + NADH + H^+$$

Succinate produced by the glyoxyllate cycle can then be converted to hexose by conversion to oxaloacetate, by the action of succinic dehydrogenase and fumarase. The oxaloacetate is converted to phosphoenolpyruvate, a reaction catalised by phosphoenolpyruvate carboxykinase,

$$\text{Oxaloacetate} + ATP \rightarrow \text{phosphoenolpyruvate} + CO_2 + ADP$$

and the phosphoenolpyruvate converted to glucose by a reversal of the reactions of the Embden-Meyerhof-Parnas pathway. Thus four molecules of acetyl-CoA will give rise to two molecules of succinate and this in turn will result in the formation of one molecule of glucose and the loss of two molecules of carbon dioxide.

The glyoxyllate cycle was first demonstrated in the bacterium *Pseudomonas* grown on two-carbon compounds, and has since been shown to operate in many micro-organisms and plant tissues. The evidence for the cycle is based on the presence of the two key enzymes, malate synthetase and isòcitrate lyase, and on the distribution of ^{14} C in organic acids, sugars and carbon dioxide when the tissue is supplied with specifically labelled [^{14} C] acetate. Malate synthetase and isocitrate lyase are found in a wide variety of plant tissues, particularly in fatty seedlings where they increase in activity during germination. Similarly the activities of aconitase and citrate synthetase increase in these tissues during germination. Incubation of castor bean endosperm tissue with [I-^{14} C] acetate or [2–^{14} C] acetate results initially in the formation of [^{14} C] malate, and subsequently the radioactivity from [I-^{14} C] acetate results in about an equal labelling of CO_2 and sucrose. In the cotyledons of germinating peanut and sunflower seedlings and castor bean endosperm, [1-^{14} C] acetate was converted to carboxyl-labelled malate and to sucrose in which the glucose moiety was labeled in the 3 and 4 carbons, while [2-^{14}]C acetate gave rise to malate labelled in the methylene carbons and to sucrose where the glucose moiety was labelled in the 1, 2, 5 and 6 carbons. These patterns or labelling of the products of acetate metabolism are consistent with the operation of a glyoxyllate cycle in these tissues.

Metabolic Functions of the Glyoxysome

The reactions of the glyoxyllate cycle and the pathway of

b -oxidation were generally thought to be associated with the mitochondria since the enzymes of these pathways were usually present in the mitochondrial fraction isolated from cell homogenates. Elegant studies by Beevers' group at Purdue University however showed that sucrose density centrifugation of crude particulate fractions of castor bean endosperm resulted in the separation of three distinct bands of particles which were identified as proplastids, mitochondria and a new particle sedimenting at a high density. Enzymes of the glyoxyllate cycle, isocitrate lyase and malate synthetase were found exclusively in this dense particle together with catalase and a large proportion of the glycollate oxidase of the gradient. On the other hand citrate synthetase and malate dehydrogenase were associated with both the mitochondrial band and the band containing the new particle, while succinic dehydrogenase, fumarase and NADH oxidase were located exclusively in the mitochondrial band with cytochrome oxidase.

The enzyme distribution indicated that the TCA cycle enzymes were located in the mitochondria while the enzymes of the glyoxyllate cycle were compartmentalised in the denser particle. These particles were therefore called glyoxysomes. The isolated glyoxysomes were found to be organelles with a single unit membrane bounding a finely granular matrix. Similar structures were recognised in electron micrographs of intact castor bean endosperm tissue indicating that the isolated organelles were not artefacts of the isolation and centrifugation processes. Since the finding of glyoxysomes in endosperm tissue, they have been reported to be the site of the glyoxyllate cycle in the cotyledons of a number of fatstoring seeds including those of watermelon, sunflower, peanut and cucumber. Microbodies containing catalase and enzymes of the glyoxyllate cycle have also been found in yeast and although these have been called peroxisomes they clearly have the enzyme complements of glyoxysomes.

In addition to glyoxyllate cycle activity, the glyoxysomes were shown to be the site of *b* -oxidation in castor bean endosperm. Glyoxysomes isolated from this tissue oxidised palmitoyl-CoA to acetyl-CoA with a concomitant reduction of NAD^+ and uptake of oxygen. Addition of [^{14} C] oxaloacetate during this reaction resulted in the formation of [^{14} C] citrate and [^{14} C] malate from palmityl-CoA indicating that the acetyl-CoA produced by the *b* -oxidation process was consumed in the glyoxyllate cycle also located in the organelle. Similarly, ricinoleate, linoleate and palmitate were oxidised by glyoxysomes of castor bean with the formation of acetyl-CoA, and three enzymes of the *b* -oxidation complex, crotonase, *b* -ketothiolase, and *b* -hydroxyacyl dehydrogenase were found to be located specifically in the organelle. The activity of the *b* -oxidation complex in this tissue during the germination of the seed was found to parallel the increase in activity of the glyoxyllate cycle enzymes indicating that an integrated system for lipid utilisation develops together with the formation of the glyoxysome.

The principal pathways of gluconeogenesis are therefore compartmentalised in a single organelle, the glyoxysome. The complete pathway of gluconeogenesis however involves at least three organelles, the spherosome, the glyoxysome and the mitochondrion. The reactions are initiated by a hydrolysis of lipids in the spherosome by the action of lipase, and the glycerol and fatty acids produced diffuse out of the organelle. Glycerol is utilised directly by the EMP pathway in the cytosol and contributes to sucrose synthesis while the fatty acids diffuse into glyoxysomes which are located near the spherosome. In the glyoxysome the fatty acid is oxidised in the *b* -oxidation pathway and the acetyl-CoA released is converted to succinate by the action of the glyoxyllate cycle enzymes located in the organelle. Although the *b* -oxidation pathway is reversible, the equilibrium is presumably maintained in a catabolic direction by the removal of peroxide, produced in the oxidation step of the pathway, by the catalase present in the glyoxysome. Reduced NAD^+ produced in these reactions is probably oxidised in the mitochondria. Succinate produced in the glyoxysome is further metabolised to oxalacetate and finally to phosphoenolpyruvate in the mitochondria since succinic dehydrogenase and fumarate are not component enzymes of the glyoxyllate cycle and are absent from glyoxysomes.

This compartmentalisation of the *b* -oxidation complex and the glyoxyllate cycle together in an organelle discrete from the mitochondrion is probably the reason why acetyl-CoA is utilised in gluconeogenesis in plant tissues rather than being oxidised to carbon dioxide and water as in animal tissues. Free acetyl-CoA-is presumably not released from the glyoxysome and made available for oxidation in the mitochondrion, while consumption of succinate, the final product of the glyoxyllate cycle, by the mitochondrion, would not stimulate the rate of oxidation in the TCA cycle since this can only occur by increasing the supply of acetyl-CoA. Some measure of control of competing metabolic pathways is therefore achieved by separation of these reactions within different organelles.

Peroxisomes

Glycollic acid is produced in large amounts in the chloroplast, as a by-product of the reactions of carbon dioxide fixation. The formation of glycollate has been proposed to occur by the oxidation of ribulose-1,5-bisphosphate (RBP) by molecular oxygen, a reaction which would produce a two carbon fragment of phosphoglycollic acid and a three carbon fragment of phospholgyceric acid (PGA) instead of two molecules of PGA resulting from the normal carboxylation of RBP by CO_2. It has been found that RBP carboxylase acts as an oxygenase in the presence of molecular oxygen and that phosphoglycollate and PGA are produced in this reaction *in vitro*. Thus the enzyme RBP carboxylase can act as an oxygenase or as a carboxylase and the formation of phosphoglycollic acid is favoured by high partial pressures of oxygen. Glycollate is produced from phosphoglycollate by the action of phosphoglycollate phosphatase, a chloroplast enzyme. Glycollate is also produced, *in vitro,* from fructose-6-phosphate by the action of the chloroplast enzyme, transketolase, which may be due to the oxidation, by hydrogen peroxide, of the glycolaldehyde-thiamine pyrophosphate, an intermediate complex in this enzyme reaction. Glycollate is released from the chloroplast into the cytosol, where it is further metabolised in a specific metabolic pathway, the initial reactions of this pathway being located in the peroxisome.

Glycollate Pathway

The pathway of glycollate metabolism in leaves has been elucidated by infiltrating excised leaves or leaf discs with ^{14}C-labelled glycollate or other intermediates of the pathway. Glycollate is converted rapidly to glyoxyllate which may be oxidised non-enzymatically to carbon dioxide and formic acid, in the presence of hydrogen peroxide. In this reaction the carbon dioxide is derived from the carboxyl carbon of glycollate and the formate from the methyl carbon. However the presence of catalase, which breaks down peroxide, is thought to preclude such a total degradation of glyoxyllate in the peroxisome and the supply of ^{14}C-labelled glycollate or glycoxyllate to leaf tissue in the light has been found to give rise initially to labelled glycine and serine and subsequently to labelled glyceric acid hexoses and sucrose.

Infiltration of [2-^{14} C] glycollate into leaves in the light was found to give [2-^{14} C] glycine but serine was found to be labelled in the 2 and 3 carbons. From this evidence it was concluded that two molecules of glycine give rise to one molecule of serine with a loss of one molecule of carbon dioxide; the 1 and 2 carbons of serine are derived from carbons 1 and 2 of glycine respectively, while the 3 carbon of serine is derived from the 2 carbon of glycine and carbon dioxide arises from carbon 1 of glycine. In wheat leaves in light [3-^{14} C] serine was converted to [3-^{14} C] glycerate and this was incorporated into hexose presumably by the reactions of the Embden-Meyerhof-Parras pathway. This pathway of hexose formation from glycollate has been confirmed by the finding that [2-^{14} C] glycollate gives rise to glycose labelled in the 1, 2, 5 and 6 carbons and [3-^{14} C] serine to glucose labelled in the 1 and 6 carbons.

The glucollate pathway is therefore gluconeogenic in light. The conversion of glyceric acid derived from glycollate, to glucose and sucrose is inhibited by DCMU an inhibitor of Photosystem II. In the dark, supplied [^{14} C]-glycollate is converted into TCA cycle acids rather than sugars which may be due to an oxidation of pyruvate derived from glyceric acid, or may result from a direct conversion of glyoxyllate to malate.

Glycollate produced during photosynthesis in the presence of $^{14}CO_2$ is usually found to be uniformly labelled i.e. both carbons have the same specific activity. This distribution of radioactivity would be expected if the glycollate was derived from carbons 1 and 2 of ribulose-1,5-bisphosphate. Consequently all the compounds arising from glycollate are uniformly labelled. Serine, in particular, has been found to be uniformly labelled while phosphoglyceric acid was predominately carboxyl labelled, indicating that serine was produced from glycollate rather than directly from phosphoglyceric acid formed in CO_2 -fixation. The uniformly labelled glyceric acid formed in this pathway in turn produces uniformly labelled hexoses instead of the 3,4-^{14} C-hexoses resulting from incorporation of ^{14} C from the photosynthetic carbon cycle.

The operation of the glycollate pathway in leaves has been shown by tracer experiments but it has been confirmed by the detection of enzymes in leaves which are necessary for some reactions of the pathway. The initial reaction, the oxidation of glycollate to glyoxyllate is catalysed by glycollate oxidase, an enzyme first isolated by Zelitch and Ochoa. It has FMN as the prosthetic group and utilises molecular oxygen as the electron acceptor. The enzyme is competitively inhibited by bisulphite addition compounds of aldehydes, *a* -hydroxy-sulphonates, which have the general structure R-CHOH-SO_3 H and are therefore structural analogues of glycollate. The most commonly used *a* -hydroxysulphonate is *a* -hydroxypyridylmethane sulphonate (HPMS) and treatment of leaf discs or infiltration of excised leaves with this compound results in the accumulation of glycollic acid while having no effect upon the rate of CO_2 -fixation. Experiments of this type have shown that 40 to 70: of the carbon fixed in photosynthesis will accumulate as glycollate in tissues treated with HPMS and these results have been interpreted as indicating that a large fraction of the carbon fixed in photosynthesis is metabolised by this pathway.

Transaminases are present in leaves which catalyse two steps of the pathway: a glutamate-glyoxyllate transaminase catalysing the formation of glycine is widespread in plant tissues, and a serine-pyruvate aminotransferase catalysing the formation of serine to hydroxypyruvate is found in leaves. The conversion of glycine to serine is catalysed by serine hydroxymethyltransferase which has been found in pea and wheat leaves. This reaction is inhibited by isonicotinyl hydrazide, and treatment of leaf tissue with this compound during photosynthesis in $^{14}CO_2$ causes an accumulation of [^{14}C] glycine and [^{14}C] glycollate and a decrease in the incorporation of ^{14} C into glucose. Use of this inhibitor thus provides additional evidence for the operation of the pathway.

Two types of glyoxyllate reductase have been found to occur in leaves. One, an NADP-linked enzyme, is thought to be located in the chloroplast, while a second NAD-linked enzyme is present in the cytoplasm and is referred to as hydroxypyruvate reductase since the enzyme isolated from some sources has a higher activity to hydroxypyruvate than to glyoxyllate. The enzyme appears to catalyse the reduction of hydroxypyruvate to glycerate in the glycollate pathway.

Photorespiration

The oxidation of glycollate in the peroxisome is accompanied by a consumption of oxygen and ultimately results in the release of carbon dioxide. The net result is a respiratory gas exchange where the substrate of respiration is glycollate rather than glucose. Since glycollate is only formed in light this respiration is light-dependent and is called *photorespiration.*

The direct measurement of oxygen uptake during a net photosynthetic release of oxygen, or CO_2 release during net photosynthetic CO_2 fixation is impossible. However, indirect methods have demonstrated that photorespiration occurs in plants and that the rate of CO_2 loss in light is higher than that in the dark. Measurements of photorespiration rates vary considerably with the assay method used and can only be considered as approximations of the magnitude of the process. Nevertheless, it is becoming clear that the process of photorespiration has great importance in decreasing the rate of net photosynthesis in many plants.

Photorespiration can be detected by measuring the flux, that is the simultaneous uptake and release, of oxygen or carbon dioxide of photosynthesizing tissue by using isotopic tracer methods. The uptake of $^{18}O_2$ by leaves during photosynthesis has been detected but experiments using $^{18}O_2$ have generally given equivocal results which have been difficult to interpret.

The measurement of carbon dioxide flux during photosynthesis is much more convenient and depends upon the accurate measurement of carbon dioxide concentration in the atmosphere using an infra-red gas analyser and a simultaneous measurement of the total activity of supplied $^{14}CO_2$ with an ion counter or Geiger-Müller counter. When a plant is placed in a closed system in light, there is a rapid uptake of carbon dioxide and the CO_2 concentration of the atmosphere around the plant decreases to a concentration at which the uptake of CO_2 exactly balances the output of CO_2 by the plant.

This concentration of carbon dioxide is called the CO_2 *compensation point* of the plant and is usually measured in parts per million (ppm) of CO_2 in air. Plants such as tobacco, sunflower and wheat have compensation points, ranging from 35 to 100 ppm indicating a marked loss of CO_2 (i.e. photorespiration) during photosynthesis, but others such as maize and sugarcane have compensation points of 3 to 10 ppm indicating a low loss of CO_2 or a low photorespiration rate. If photosynthetic carbon fixation is similarly measured in a closed-system but in an atmosphere containing $^{14}CO_2$, the CO_2 arising from, the plant by photorespiration will, over short time periods, be $^{12}CO_2$. Thus, while a decrease in the CO_2 *concentration* of the atmosphere around the plant will occur, the *radioactivity* of $^{14}CO_2$ in the atmosphere will appear to decrease at a faster rate because of the efflux of unlabelled CO_2 from the plant, and continues to decrease even after the compensation point is reached.

In a plant which is photorespiring, therefore, a decrease in the *specific radioactivity* of $^{14}CO_2$ will occur, while in a plant with no photorespiration the uptake of $^{14}CO_2$ will not be accompanied by an efflux of $^{12}CO_2$ and the specific radioactivity of the $^{14}CO_2$ in the atmosphere will remain constant. In the dark the specific radioactivity of the $^{14}CO_2$ decreases with the efflux of $^{12}CO_2$ and the rate of decrease of this activity is a measure of dark respiration. The rate of photorespiration as measured by these methods has been shown to be 1.5 to 2.5 times that of dark respiration, while in plants with low photorespiration, such as maize, loss of CO_2 in light is only a fraction of the rate of CO_2 loss in the dark. This method clearly indicates that photorespiration occurs in some species while it is absent in others.

Another method which has been used to detect differences in the rate of photorespiration and dark respiration was devised by Zelitch. In this method leaf discs are allowed to fix $^{14}CO_2$ for a period of 45 to 60 minutes. The remaining $^{14}CO_2$ is then quickly flushed out of the closed system, and the release of $^{14}CO_2$ from the tissue into CO_2 -free air is measured over short periods in the light and the dark. The method is based on the assumptions that the rate of $^{14}CO_2$ loss is a measure of total CO_2 loss i.e. that the specific radioactivity of the CO_2 evolved remains constant over short time periods and that low CO_2 tensions have no effect on the loss of CO_2. These assumptions may not be valid for all photosynthetic tissues but within these limitations the method is a very rapid and sensitive means for detecting photorespiration. Photorespiratory loss of CO_2 in tobacco has been shown by this method to be 2 to 5 times higher than in the dark while CO_2 loss from maize is not detectable. This method has been similarly used to detect photorespiration in other species.

Photorespiration is not simply a stimulation of dark respiration since the two process of photorespiration and dark respiration respond differently to changes in oxygen concentration. Dark respiration of leaves has an optimal rate at about 2% oxygen in air and any increase in the O_2 concentration up to 100% does not increase the rate. Net photosynthesis however is inhibited by oxygen, a phenomenon called the *Warburg effect*, and this is attributable to an increase in the rate of photorespiration rather than to an inhibition of photosynthesis *per se* . A reduction in the oxygen concentration in the atmosphere around a leaf from the ambient 21% O_2 of air has been shown to lower the compensation point and increase the rate of net photosynthesis, while an increase in concentration raises the compensation point and lowers net photosynthesis.

Plants with photorespiration have been found to evolve CO_2 at a high rate when they are transferred to darkness after a period of photosynthesis. This post-illumination CO_2 burst lasts for a few minutes before a steady dark respiration is established and the magnitude of the burst has been found to depend on the light intensity and oxygen concentration during the previous period of photosynthesis: the CO_2 -burst increases with an increase in light intensity and decreases with a decrease in O_2 concentration. This burst has been explained as being due to the continued slow oxidation, in the dark, of a product produced in photosynthesis after the uptake of CO_2 by photosynthesis has stopped. Photorespiration therefore appears to have different characteristics than dark respiration and to have a different substrate.

The substrate for photorespiration is thought to be glycollic acid. Glycollic acid production in leaves and in chloroplasts is stimulated by an increase in oxygen concentration as is photorespiration, while the inhibition of glycollate oxidation by HPMS has been found to lower the rate of photorespiration in leaf discs to that of dark respiration. Photorespiratory loss of CO_2 is also stimulated by the infiltration of leaves with glycollic acid while acetate has no effect on this rate. Thus the oxidation of glycollate, mediated by the leaf peroxisomes results in a photorespiratory loss of carbon dioxide by the leaf.

Plants with low compensation points are said to lack photorespiration. These plants are C_4 plants, that is, the primary carboxylation step is catalysed by phosphoenolpyruvate carboxylase rather than by RBP carboxylase. This efficient fixation of CO_2 has been postulated to preclude the formation of glycollate, but maize for example, has been shown to possess a glycollate pathway and peroxisomes occur in the mesophyll cells of the leaf. Presumably in such plants glycollate may be formed and oxidised, but the efficiency of refixation of the CO_2 resulting from glycollate oxidation is such that no CO_2 is released from the plant and hence no photorespiration is detected.

Estimates from various plant species suggest that from 15 to 40% of the carbon fixed in photosynthesis is lost by photorespiration. There is no unequivocal evidence to show that energy in the form of ATP is recovered during the oxidation of glycollate and the process appears to be a wasteful one in terms of energy conservation. An important question therefore arises: does photorespiration serve a useful function or is it simply an inevitable consequence of photosynthesis being carried on in an atmosphere of 21% oxygen? It has been suggested that photorespiration might act as a 'safety valve' for the plant, in that, under conditions of high light intensity and low carbon dioxide concentration, the oxidation of glycollate would consume both excess reduced $NADP^+$ and excess oxygen, which would serve to protect the chloroplast from photo-oxidative damage.

If the process of photorespiration imposes limitations on the growth of plants, it would be expected that there would have been some selective pressure to eliminate it by natural selection during the evolution of the higher plants. The occurrence of C_4 plants, which have low photorespiration, may represent such an evolutionary step to correct for the presence of photorespiration by the development of an efficient CO_2-fixation mechanism. Since variations in the rate of photorespiration occur within a single species it may be possible to select artificially for low photorespiration in crop plants and consequently increase net photosynthetic productivity and crop yield.

Ontogeny and Turnover of Microbodies

The development of microbodies in plant cells is a difficult process to follow since they usually occur in small numbers and are difficult to characterise enzymatically. Small, membrane-bound organelles occur in close proximity to the endoplasmic reticulum but these do not always give a positive reaction to DAB, the usual criterion for the presence of catalase. Nevertheless catalase is invariably the first enzyme to be detected in microbodies and it appears to be the most active enzyme in the developing organelle. Studies of microbody development have been carried out principally on tissues having large populations of microbodies such as the endosperm and cotyledons of fatty seedlings.

In the endosperm of fat-storing seeds, particularly of the castor bean, the process of germination results in the development of large numbers of microbodies which reach numbers per cell twice that of mitochondria. This increase in number is correlated with increases in the activities of isocitrate lyase and malate synthetase in the whole tissue. It has been suggested by several investigators that microbodies develop from the endoplasmic reticulum by a budding process.

In castor bean the formation of microbodies has been linked directly to the endoplasmic reticulum and connections between the cisternae of the rough endoplasmic reticulum and developing microbodies have been clearly demonstrated. This connection does not appear to persist during the whole course of organelle development although they increase seven-fold in size and increase in density suggesting a continuous addition of newly synthesized protein. The depletion of lipid in the endosperm or cotyledons of fatty seedlings is accompanied by a decrease in glyoxysomal enzymes which is attributable to a decrease in the number of glyoxysomes as measured by enzyme activity and total protein. It is not clear how these microbodies are destroyed but some at least become enclosed in autophagic vacuoles and are digested.

Increases in peroxisomal enzymes have been detected in germinating seedlings and etiolated seedlings on exposure to light and some of these changes have been correlated with the size and structure of microbodies. The development of microbodies in, leaves of wheat seedlings in light involves distinct changes in the biochemical and physical characteristics of the organelle. During the first few days of germination an increase occurred in the density of catalasecontaining particles from 1.17 to 1.24 g cm^{-3} on separation on sucrose gradients, which was paralleled by an increase in catalase, hydroxypyruvate reductase and glycollate oxidase.

In etiolated leaves small microbodies occur which are rich in catalase but low in other peroxisomal enzymes. These bodies have not been unequivocally characterised as peroxisomes because of the difficulty of isolation. Upon exposure to light there is an activation of glycollate oxidase and hydroxypyruvate reductase, but these changes in enzyme activity have not been correlated with an increase in size and number of peroxisomes. The activity of glycollate oxidase is stimulated by exposure of etiolated tissue to red light and this activation is reversed by -far-red light. These effects probably reflect a stimulation of leaf development rather than a specific phytochrome involvement in peroxisome development. However an increase in glycollate oxidase occurs under continuous far-red light in mustard seedlings where chloroplast development is inhibited, suggesting that substrate activation is not involved in this increase of enzyme activity. Similar increases in the activities of glycollate oxidase, and hydroxypyruvate reductase have been found in cotyledons of germinating seedlings on exposure to light.

In germinating seeds and in greening leaves the microbodies have easily identifiable functions, those of glyoxysomes or peroxisomes, and only one type of microbody occurs in each tissue. However an interesting situation occurs in some seeds of the *Cucurbitacea.* In cucumber and watermelon for example, lipid is the main storage material in the seed and during germination there is an increase in activity of glyoxysomal enzymes in the cotyledon. During normal development, the cotyledons become exposed to light and turn green. The glyoxysomal activity of the microbodies, as measured by the activity of malate synthetase and isocitrate lyase, decreases while the peroxisomal activity, as measured by glycollate oxidase activity, increases.

In one organ therefore in response to environmental stimulus there is a transition from lipid breakdown and gluconeogenesis, to glycollate metabolism, which is reflected in a change in the microbody population from a glyoxysomal to a peroxisomal function. Since the two microbodies are morphologically indistinguishable the question arises as to whether the structure of the glyoxysome is retained and there is a selective replacement of enzymes to give it a peroxisomal function or whether there is an autolysis of the glyoxysome and a replacement of it by a newly synthesized peroxisome.

The changes in microbody enzyme complements which occur during seedling development have been correlated with changes in the fine structure of the microbodies. It has been reported that in cucumber seedlings the decrease in glyoxysomal enzyme activity is not accompanied by a decrease in the number of microbodies as determined by electron microscopy. The subsequent exposure of the cotyledons to light caused a great increase in the activity of peroxisomal enzymes, particularly that of glycollate oxidase, while there was not a concomitant increase in the number of microbodies. No evidence was found for the degradation of microbodies during the loss of glyoxysomal enzyme activity as has been reported in castor bean. These studies suggest that there is a continuity of microbody structure in the transition from a glyoxysomal to a peroxisomal function. On the other hand, in developing watermelon seedlings, the decrease in the total activity of glyoxysomal enzymes in isolated microbodies which occurred after the depletion of lipid, was accompanied by a decrease in the total protein of these microbodies. This was interpreted as indicating a destruction of glyoxysomes in the tissue.

Normally the peroxisomal activities of the microbody increase as glyoxysomal activity decreases, but in this tissue the decrease in glyoxysomal activity of the microbodies could be separated from the increase in the peroxisomal activity. Brief exposure of dark-grown cotyledons to light at an early stage of development stimulated the activities of peroxisomal enzymes, while this did not affect the rate of decline of the activity of glyoxysomal enzymes. These results suggest that the formation of peroxisomes can occur while active glyoxysomes are present and that two populations of microbody can occur in the tissue at the same time. The apparent contradiction in the results from these two tissues has still to be resolved, and it is still not clear whether the transition from a glyoxysomal function to a peroxisomal function in the microbodies of these tissues is due to enzyme replacement in a single microbody or to synthesis of a new microbody.

Algal Microbodies

The microbodies of photosynthetic higher plants have distinct physiological roles either as peroxisomes or glyoxysomes which are active at different stages of the life cycle of the plant. In the algae however their role is not quite as clear since both functions of the microbody, i.e. glycollate oxidation and gluconeogenesis, may be carried on in a single cell. Microbody-like organelles have been found in a wide range of algae, and they are most numerous in algal cells grown on two-carbon compounds. Catalase has been detected cytochemically by DAB staining *in situ* but has not been shown to be universally present in algal microbodies.

Algae grown autotrophically, that is, in light on carbon dioxide, produce glycollic acid which is metabolised by a pathway similar to that of higher plant leaves. Glycollate oxidase however is not present in algae, with the possible exception of some members of the Zygnematales, Ulotrichales and Charophyceae but the oxidation of glycollate is catalysed by glycollate dehydrogenase. This enzyme is not FMN dependent and does not couple to molecular oxygen so that hydrogen peroxide is not formed during glycollate oxidation. This may to some extent account for the low activities of catalase reported in many algae. The natural electron acceptor for glycollate dehydrogenase is as yet unknown.

While the presence of a glycollate pathway has been established in green algae, the extent to which glycollate is metabolised by this pathway during photosynthesis is not known. In contrast to higher plant cells, serine formed during photosynthesis in $^{14}CO_2$ in several algae is carboxyl labelled indicating its formation directly from PGA rather than by the glycollate pathway. There is also a lower rate of CO_2 loss by photorespiration in algae than in higher plants suggesting perhaps that little glycollate is formed in algal photosynthesis. Glycollate is excreted by algal cells and this has been interpreted to indicate that the algae have a low rate of glycollate metabolism. This can now be discounted however since little glycollate is excreted during steady-state photosynthesis and all the glycollate formed appears to be metabolised in the algal cell.

Many unicellular algae can be grown in the dark on two-carbon compounds such as acetate or ethanol as sources of both energy and carbon for growth. *Chlorella* grown in the dark on acetate, incorporates [^{14}C]-acetate into protein and carbohydrate without prior degradation while simultaneously about half of the acetate taken up is oxidised to CO_2 in a manner consistent with the operation of the TCA cycle. These cells have high levels of malate synthetase and isocitrate lyase, thereby allowing the formation, from acetate, of four-carbon compounds which can form precursors of amino acids and hexoses. Autotrophically-grown cells however have little isocitrate lyase activity and [^{14}C]-acetate supplied to these cells in the dark is oxidised to CO_2. Furthermore the activity of isocitrate lyase increases when autotrophically-grown cells are supplied with acetate in the dark and no cell division takes place during this time. Similar increases in malate synthetase and isocitrate lyase in response to growth on two-carbon compounds have been reported in *Euglena gracilis.*

The enzymes of both the glycollate pathway and the glyoxyllate cycle have therefore been found in several algae and the question arises as to whether these enzymes are compartmentalised in a microbody. The limited number of studies which have been done suggest, on the basis of their enzyme activities, that microbodies isolated from algae have either a glyoxysomal function or are of the non-specialised type often present in non-green plant tissues.

In *Euglena gracilis* grown on ethanol the clyoxylate cycle enzymes malate synthetase and isocitrate lyase together with glycollate dehydrogenase and glyoxyllate reductase were detected in a particulate fraction distinct from mitochondria and, while no catalase activity was detected, the enzyme complement indicates a microbody with a glyoxysomal function. However in another flagellate, *Chlorogonium elongatum,* grown either on acetate in the dark or autotrophically in the light, a catalase-containing microbody was found which had neither glyoxysomal nor peroxisomal enzymes associated with it. In the acetate-grown cells isocitrate lyase and malate synthetase were found to be located in the cytosol, while in autotrophically-grown cells, glycollate dehydrogenase, hydroxypyruvate reductase and glyoxyllate-glutamate aminotransferase were$_2$ found to be localised in a mitochondrial fraction together with cytochrome oxidase, and malate and isocitrate dehydrogenases.

It is apparent that algal microbodies do not have a peroxisomal function and that this is correlated with the oxidation of glycollate in these cells by glycollate dehydrogenase, thus eliminating the requirement for the compartmentalisation of glycollate oxidation with catalase. However, microbodies have been isolated from only a few algae primarily because vigorous methods are required to break algal cell walls and microbody isolation is therefore difficult to achieve. It remains to be seen whether peroxisomes occur in algae such as *Nitella* and *Spirogyra* which have relatively high levels of catalase and also have glycollate oxidase rather than glycollate dehydrogenase.

MICROTUBULES

Microtubules were first described in spermatozoids of the moss, *Sphagnum*, by Irene Manton in 1957. Since the introduction, six years later, of glutaraldehyde as a fixative for electron microscopy, their presence has been revealed in a very wide variety of plant and animal cells. They have not been observed in prokaryotes.

These organelles are associated with several processes in the repertoire of movements exhibited by the eukaryotic cell, including; movements of individual cells by flagellar motion or axopod extension; movement of components within cells, as in mitosis, axoplasmic flow or transport of certain pigment granules; and morphogenetic movements involving the generation and maintenance of cell shape. In the case of plant cells, in addition to chromosome movements during cell division, the presence of microtubules has been correlated with definition of the plane of cell cleavage, formation of the cell plate and determination of cell wall architecture.

In certain cell types, microtubules are seen only during mitosis and meiosis, suggesting that they were involved originally in these processes alone, and only later did they come to be used for extranuclear functions in advanced eukaryotic organisms. Indeed, it has been suggested that acquisition of the microtubule was a major factor in permitting the evolution of the eukaryote cell.

Under the electron microscope, microtubules appear as long, unbranched cylindrical structures with a diameter of 22–25 nm. In transection, they show an electron-lucent core, *ca.* 12 nm in diameter, and a wall of, usually, 13 electron-dense subunits. The subunits are 4–5 nm wide and are organised along the long axis of the microtubule into protofilaments. There is an axial displacement of subunits between adjacent protofilaments, resulting in a visible pitch relative to the long axis of the tubule.

Microtubules are usually separated from each other or from adjacent cellular components by a clear space of 10–40 nm. Also, in longitudinal view, an electron-lucent zone is often observed along the sides of the microtubule. These observations suggest that each tubule may be surrounded by a specialised region or layer of material. In certain systems, microtubules show 'arms' projecting from the walls of the tubules at regular intervals along their length.

Such lateral projections act, in some instances, as cross-bridges between adjacent microtubules. Alternatively, projections can extend from microtubules to adjacent membranes such as the plasmalemma, the nuclear envelope, the endoplasmic reticulum or associated vesicles.

Drug Interaction

The responses of certain types of microtubule to anti-mitotic agents, together with their apparent repeating substructure, thus gave rise to the concept of the microtubule as an assembled polymer of soluble subunits. Early investigations into the nature of the subunit were hampered by the absence of a method or an assay for the recognition of a soluble microtubule component, even when such enriched sources as flagella or the mitotic apparatus were examined. This problem was overcome in the mid-1960's when it was postulated that the microtubule system was the direct target of the anti-mitotic drug, colchicine, and that the subunit might therefore be recognisable as a drug receptor.

Preliminary experiments with human carcinoma cells in culture established that, at the low levels of colchicine which depolymerised microtubules, there were no effects on other areas of metabolism such as respiration or protein synthesis. Consideration of the kinetics of drug uptake suggested the presence of a single type of binding site. Work in two laboratories confirmed, using radiolabelled drug, that colchicine itself was indeed bound non-covalently to a single, soluble, protein species. Since colchicine-binding could be assayed quantitatively, using gel filtration or ion exchange procedures, the reaction assumed great importance in the subsequent isolation and characterisation of the subunit.

It was quickly established that colchicine is bound in a stoichiometric 1:1 relationship to a dimeric protein with a molecular weight of 120,000. Denaturation to the protomers results in loss of binding. The binding affinity is high, of the order of 2×10^6 litres $mole^{-1}$, and it is important to note that there is no binding at all of the colchicine isomer, lumicolchicine. Although the protein can be stored in the frozen state, an outstanding feature of the binding site is its instability in solution, where it shows first order decay kinetics and a half-life of the order of hours even under optimal conditions of pH, temperature and ionic environment. Inclusion in the medium of the nucleotide, GTP, or the alkaloid, vinblastine, stabilises and binding activity to some extent.

Evidence that the colchicine binding component is the microtubule subunit was initially circumstantial. High levels of colchicine binding protein were obtained from material rich in microtubules, e.g. dividing cells, cilia and brain cells. The physical characteristics of the binding protein with regard to its molecular weight and dimeric nature were similar to those of the major protein species in flagella and cilia. Evidence of a more direct nature was obtained through ultrastructural observations of the solubilisation of the central pair nucleotide, GTP, or the alkaloid, vinblastine, stabilises the binding activity to binding activity.

The binding site for colchicine seems to be masked when the subunits are assembled into microtubules. This, together with the instability of the site, was responsible for initial lack of success in obtaining colchicine-binding activity from stable outer doublet microtubules of flagella. However, the presence of the colchicine site in such subunits has been demonstrated using appropriate solubilisation procedures. Thus, the colchicine binding protein has been obtained from a variety of types and sources of microtubule. It is now generally accepted that labile microtubules are depolymerised by the binding of colchicine to subunits in the soluble pool, with consequent disruption of the equilibrium.

The colchicine binding moieties isolated from diverse sources of microtubules display similar physical and chemical properties. The name 'tubulin' has been applied to this dimeric, functional subunit of microtubules. Tubulin is an acidic, globular protein with a sedimentation coefficient of 6s and an agreed molecular weight of 110–120,000, whose behaviour approximates that of a prolate ellipsoid with an axial ratio of 5:7. The amino acid composition prompted initial comparison with actin, but subsequent comparable peptide maps of both proteins argued against any homology between tubulin and actin.

Treatment of tubulin with chaotropic agents such as urea or guanidine. HCl results in a reduction of the apparent molecular weight to 50–60,000. The two protomeric polypeptides can be resolved by appropriate electrophoretic analysis and have been named *a* and *b* tubulin, the more mobile electrophoretic component being *b* tubulin. A number of chemically reactive sites have been distinguished on the tubulin heterodimer. Two sites are known to be involved in drug interaction, a site for colchine and podophyllotoxin, and a separate site for vinblastine. There are also sites which bind endogenous

ligands, including a site where GTP (guanosine triphosphate) is tightly and non-exchangeably bound and another site to which GTP is bound in such a way as to allow exchange with exogenously added GTP. Dephosphorylation of the terminal phosphate of GTP and transphosphorylation between the two sites have been suggested to play a role in microtubule assembly. In addition, other sites may be involved in interactions with Ca^{2+} and Mg^{2+} ions. The relationship of all these sites to the protein-protein contact sites necessarily involved in the vertical and lateral bonding of the subunits in the assembled tubule is unclear.

Recent examinations of plant preparations for a tubulin-like moiety with characteristic colchicine-binding activity have met with limited success. Such activity has either been undetectable or present only to a low degree. The stability and specificity of the binding render it unlikely that all such activity is attributable to a plant tubulin.

Polymerisation

Since the behaviour of microtubules *in vivo* obviously involves their strictly regulated assembly-disassembly, a primary aim throughout the biochemical investigations was to develop a system in which polymerisation could be studied *in vitro*. Early studies in this area did indeed demonstrate the presence of ordered structures in tubulin preparations. However, the relevance of such systems to the situation *in vivo* was considered doubtful, either because the process did not show characteristic kinetic responses to colchicine and temperature, or because the structures did not look like microtubules. It was therefore a notable advanace when authentic microtubules from rat brain, sensitive to colchicine and temperature, were first assembled *in vitro*, the crucial factors being inclusion in the medium of GTP and the calcium-chelating agent, EGTA (ethylene-glycol-amino ethylether tetracetic acid).

It was subsequently reported that polymerisation, using pig brain tubulin, also requires nucleation sites which are in the form of discs (29 nm diameter, 17 nm lumen). These successful polymerisation studies finally confirmed that the colchicine binding protein is the microtubule subunit. In addition, they have provided a means of demonstrating the probable universality of the tubulin moiety, via co-polymerisation of tubulin from such disparate sources as *Chlamydomonas* flagella and pig brain.

The ability to polymerise microtubules *in vitro* has initiated several exciting new avenues of research. In the first place, it has provided a method for isolating *microtubule* protein, rather than tubulin, the drug receptor. Cycles of centrifugation at 37°C and 0°C allow microtubules to be isolated with a high degree of purity. Both the ultrastructure and the chemical composition of such preparations have been examined. Electrophoretic analyses of purified microtubules by several laboratories has confirmed the presence of more than ten other proteins, including some with a molecular weight some with a molecular weight greater than 200,000. The roles of these proteins are the subjects of intense investigation. Some seem to be involved in the regulation of assembly. Others, with a molecular weight close to that of dynein may be involved in the functioning of microtubules.

Secondly, *in vitro* polymerisation has made the assembly process itself amenable to experimental study. The initial test of biological relevance seems settled, with endothermic, reversible, polymerisation system responding to many factors known to regulate *in vivo* assembly of microtubules. The questions of nucleation centres and morphopoeitic proteins have already been mentioned. In addition microtubules have been successfully polymerised *in vitro* on cellular components such as asters, basal bodies and kinetochores. The pathway of polymerisation is being deduced from combined biochemical and ultra-structural examination of experimentally induced intermediates. A commonly observed intermediate is a ring structure of tubulin, varying in dimensions, and complexity with different investigators. Whether the ring structure serves as a nucleation centre for stack-wise growth of microtubules, or whether it uncurls to form protofilaments which then associate laterally, is unclear.

Functional Roles in Plant Cell

Microtubules appear to be involved in several functional roles within the plant cell. For convenience, these can be grouped under five headings: (1) Cytoskeletal role; generation and maintenance of cell shape (2) Cell wall architecture; cell differentiation (3) Intracellular transport (4) Chromosome movements; cell plate formation (5) Cell motility; cilia and flagella. In this section, we shall consider the involvement of microtubules in these phenomena, with emphasis on possible operative mechanisms, and drawing, and drawing, where necessary, on information obtained from studies of animal cells.

Cytoskeletal Role

Despite the forces of surface tension, animal and some lower plant cells can maintain a non-spherical shape in the absence of a rigid cell wall. Numerous electron microscope studies have revealed that microtubules are present in large numbers and with distinctive orientations in such asymmetric cells or cell extentions. Correlations between cell asymmetry and microtubules have been extensively studied in animal cells, for example in erythrocytes, explanted nerve ganglia or neuroblastoma cells, lens epithelial cells and mesenchyme cells of echinoderm embryos. When these cells are treated with microtubule depolymerising agents, they assume spherical shapes.

The unicellular alga, *Ochromonas,* lacks a cell wall or pellicle, yet has a characteristic pear-shape, with a narrow tail or rhizoplast, and a bulbous anterior bearing a small projection, the beak. Two sets of microtubules are involved in maintaining this shape. The plasma membrane of the main cell body is underlain by a series of curved microtubules which extend into the rhizoplast. The other set of microtubules is associated with the beak complex. Disassembly of the microtubules with colchicine or hydrostatic pressure leads to loss of the characteristic cell shape; removal of the depolymerising agent allows the normal shape to regenerate even in the absence of protein synthesis.

The two sets of microtubules in *Ochromonas* are differentially sensitive to colchicine and pressure. At relatively high concentrations of colchicine (10 mg ml^{-1}) or at high pressure (8,000 psi), both sets of microtubules are affected but at different rates: first, disassembly of rhizoplast tubules is correlated with disappearance of the posterior tail; this is followed by loss of beak tubules and beak asymmetry. Lower concentrations of drug (5 mg ml^{-1}) or lower pressure (6,000 psi) selectively disassemble only the rhizoplast microtubules. These results possibly indicate separate roles for each set of microtubules in determining the shape of this cell. Assuming that different tubulins are not involved, the beak and rhizoplast microtubules may represent two equilibrium systems competing for a common pool of subunits. Alternatively, the differential sensitivity of the two systems, and their sequential reassembly, may reflect characteristics of the sites which initiate microtubule polymerisation.

The protozoan, *Actinosphaerium*, has been extensively used in similar studies on the morphopoietic function of microtubules. The organism is a spherical cell bearing numerous, long, relatively stiff, protoplasmic extensions or axopods, containing a central core, the axoneme. Each axoneme is composed of several hundred axially aligned microtubules, arranged into two rows that coil about a central axis. This strict pattern does not seem to be determined by a template of initiating sites, since, during axopod growth, the microtubule array is initially disorganised and becomes progressively more geometrically perfect.

When low temperature, pressure or drugs are employed to depolymerise microtubules of *Actinosphaerium*, the axopods undergo immediate retraction. In contrast, D_2O, which stabilises microtubules, prevents the breakdown of axopods by cold or pressure. Retractions and extensions are a normal feature of individual axopods and serve to move the cell across its substratum. Thus, the cell must exert a fine control over the behaviour of microtubules in each axopod separately..

In flowering plants, the male sperm is formed from the generative cell which usually has no cell wall and lies within the pollen grain. Growth of the pollen tube is accomplished by the vegetative cell, while the generative cell adopts a spindle-shape before entering and travelling down the pollen tube. Development of the spindle shape is accompanied by the appearance of microtubules aligned parallel to the long axis of the generative cell. Treatment of the generative cell with drugs leads to a loss of microtubules and the cell adopts a spherical form.

Thus, microtubules are apparently involved in both generating and maintaining changes in cell shape. The mechanisms regulating orientation and assembly of microtubules in this role are still unknown. Microtubules in the axonemal cores of *Actinosphaerium* axopods are connected with their neighbours by cross-bridges. Whether these lateral links function in stabilising the extended structure or in orienting the microtubules has yet to be determined.

Cell Wall Architecture

During interphase in higher plant cells, cytoplasmic microtubules lie close to the cell membrane and are arranged circumferentially along the lateral walls of the cells but are randomly disposed underlying the transverse walls. In the earliest paper reporting the presence of microtubules in higher plant cells, Ledbetter and Porter noted the parallel alignment of cellulosic microfibrils of the wall and the subjacent cytoplasmic microtubules. Numerous subsequent studies have confirmed this observation, in algal cells and in higher plant cells undergoing both primary and secondary wall formation.

In the long xylem fibres of *Salix*, wall microfibrils are deposited in two different orientations simultaneously in the middle and at the extremities of the cell. Even in this situation, microtubule alignment mirrors that of the overlying microfibrils, indicating the ability of the cell to maintain two sets of microtubule-microfibril associations.

In the lorica stalk of the alga, *Poteriochromonas*, the primary wall microfibrils are arranged helically and tend to fasciate into ribbonlike fibrils, 20 nm in width. Every such wall fibril coincides precisely with a microtubule, the plasmalemma separating the two sets of linear structures.

Other examples of the association between microtubules and microfibrils are afforded by cells undergoing irregular or sculptured deposition of secondary wall material. During xylogenesis, microtubules are specifically grouped under the developing wall thickenings in vessels, and, once again, are oriented parallel to the microfibrils. Similar situations prevail during the development of wall thickenings in differentiating guard cells and during the deposition of the nacreous walls of sieve elements.

Attempts have been made to confirm the association between wall microfibrils and microtubules by treating cells with colchicine. However, plant tissues seem relatively resistant to a wide variety of drugs and, even at high concentrations of colchicine, not all cytoplasmic microtubules in plant cells are depolymerised. Also, such levels of colchicine are known to affect other cellular processes, in particular, membrane-related phenomena. Therefore, although wall microfibril orientation is often distorted in the presence of colchicine, such results should be interpreted with caution.

Cell Division

At least two types of microtubule can be distinguished in most spindles, those that traverse the spindle from pole to pole (continuous fibres), and those that connect the kinetochores of the chromosomes to the poles (chromosomal or kinetochore fibres). During separation of chromatids in anaphase, the continuous fibres lengthen, increasing the distance between the poles, while the chromosomal fibres usually—but not always—shorten, further separating the chromatids. After telophase, the continuous spindle may persist and be added to in the interzonal region by numerous additional microtubules, leading to formation of the phragmoplast and eventually, the cell plate.

Considerable changes in microtubule organisation take place during the events of a typical mitotic cycle. With the onset of division, the peripheral cytoplasmic microtubules of the interphase cell disappear. Concomitantly, a band of microtubules is formed, several layers deep, lying close to the cell wall and girdling the nucleus perpendicularly to the prospective spindle axis. This preprophase band of microtubules appears to predict the plane of the future cell plate. This has been demonstrated in divisions destined to give daughter cells of different size, for example in the asymmetric divisions of guard mother cells of wheat.

At the beginning of prophase, the number of microtubules in the preprophase band decreases, and new microtubules appear close to the nuclear membrane, this time running parallel to the prospective spindle axis. The microtubules in this so-called 'clear zone' in turn appear to furnish protein material for the spindle. It is possible that a total transformation of these extranuclear microtubules into spindle microtubules takes place.

Prior to spindle formation, the nuclear envelope either disintegrates or becomes perforated at the prospective poles, and cytoplasmic (clear zone) microtubules enter the karyoplasm at these sites. Some of these microtubules appear to establish connections with the chromosomal kinetochores. Since the number of kinetochore microtubules (kMts) increases as prometaphase proceeds, and since clear zone microtubules can be utilised only immediately after breakdown of the nuclear envelope, it is likely that *de novo* microtubule assembly also occurs at the kinetochores. Thus, it appears that the kinetochore, which varies in appearance in different organisms from amorphous to highly structured, acts as an initiating site or microtubule morganising centre (MTOC).

In prometaphase, continuous fibres also begin to form and to increase in number. There is some controversy as to whether in higher organisms the continuous microtubules pass from pole to pole. A large proportion appear to overlap in the interzone between the half-spindles, and have been termed non-kinetochore microtubules (nkMts). In lower plants, and all animal cells, with centrioles, the continuous or nkMts originate near the centrioles. These loci are represented by amorphous, electron-dense aggregates of material, and behave as another set of MTOCs. The essential feature of the centriole in mitosis seems to be in determining the spindle axis. In higher plants that lack centrioles, neither the location nor form of polar MTOCs is clear.

During metaphase, engagement of sister chromatids to opposite poles takes place *via* microtubules, followed by movement of the chromatids to the equatorial plate. At this stage, the maximum number of spindle microtubules is attained, the numbers often trebling between prometaphase and metaphase. Estimates from microtubule counts in serial sections of the metaphase spindle of the sea urchin, *Arbacia,* indicate a total microtubule length of 50,000 μm. The very large Anaphase separation of the chromatids involves direct translation of the kinetochores at almost uniform velocity (0.2–4 μm min^{-1}) towards their respective poles. Two experimentally separable processes are involved: first, a movement of the chromatids towards their poles, during which the chromosomal fibres shorten to as little as 20% of their metaphase length; secondly, an elongation of the spindle itself, during which the distance between the poles increases, thereby further separating the two groups of chromatids.

Time-lapse films of dividing *Haemanthus* endosperm cells show that many other objects, such as vesicles and granules, are commonly transported polewards during prometaphase-anaphase. Conversely, in onion root tip cells, the poleward movement of the chromosomes is accompanied by a reciprocal movement of other materials from the poles towards the mid-plate. Elements of the endoplasmic reticulum accumulate at the poles during metaphase, enter the spindle at anaphase, pass between the chromosomes moving in the opposite direction, and eventually aggregate and coalesce at the mid-plate to give rise to the transverse cell wall at telophase.

Dissolution of kMts and nkMts during telophase is accompanied by the formation of new microtubules in the interzone, leading to the formation of the phragmoplast in higher plant cells. Phragmoplast microtubules appear to be involved in the movement, alignment and fusion of ER- and dictyosome-derived vesicles to form the cell plate—yet another example of microtubule-associated intracellular transport.

Thus, distinct changes in the polarity and distribution of spindle microtubules take place during the various phases of the mitotic cycle. It should be borne strongly in mind that such changes occur under conditions of very low protein synthesis. Consequently, both assembly and functioning of the mitotic apparatus must involve regulation of microtubule polymerisation at a physico-chemical level. Interaction of microtubules with other components of the mitotic apparatus must also be regulated at this level. Several hypotheses have been proposed to explain the mechanism of mitotic movements, although little direct experimental evidence to support the models has yet been obtained.

One prevailing hypothesis to account for spindle elongation is that the poles are pushed apart by tip growth of continuous microtubules in the course of which subunits from disassembled kMTs may be utilised. Shortening of the chromosomal fibres is thought to occur by depolymerisation at the poles.

Another hypothesis based on the sliding of adjacent spindle microtubules was proposed by Mcintosh *et al.* Earlier, several workers had reported the existence of arms or cross-bridges between spindle microtubules. The basis of the sliding tubule hypothesis is that the intertubular bridges in the spindle serve as mechano-chemical elements. Sliding of two microtubules against each other is thought to be generated by the successive breaking and reforming of cross-bridge linkages, a process that is energy-consuming and mediated *via* ATPase activity of the bridges. It has been estimated that the energy released through the hydrolysis of approximately 20 ATP molecules would suffice to move an average chromosome from the metaphase plate to the pole.

A prerequisite for the McIntosh model is an ordered system of long microtubules, emanating from the poles and reaching far into the opposite half-spindles in metaphase. Microtubules from opposite poles are suggested to have antiparallel polarity with regard to the direction in which their lateral arms could produce force. The kMts in each half-spindle would also show polarity; their lateral arms would only produce force directed towards the spindle equator. If one of the long interpolar microtubules and a kMt of antiparallel polarity came into contact at anaphase, the two tubules, due to the directed activity of their arms, would push themselves in opposite directions by sliding. The model is analogous to the generally accepted model for the interaction of actin and myosin in skeletal muscle. The mechanism could cause chromosome separation, as well as spindle elongation when anti-parallel nkMts interact where they overlap.

A third hypothesis involving actin-microtubule interactions has derived from observations of actin-like thin filaments in the mitotic or meiotic spindles of both animal and plant cells. Fluorescein-labelled heavy meromyosin binds to isolated sea urchin spindles. At the ultrastructural level, microfilaments within the spindle have been decorated with heavy meromyosin, a test for actin-like proteins. Several other recent publications have confirmed the presence of actinlike filaments in dividing cells. In kangaroo rat cells, actin was only demonstrable in association with chromosomal fibres, suggesting an actinmyosin *type* interaction as the force-producing mechanism for chromosome movements, that is, for the autonomous movements of chromosomes in anaphase, as opposed to spindle elongation which may be controlled solely by microtubule assembly-disassembly.

Myosin-like proteins have not yet been located in spindles. Several workers have suggested that microtubules might serve as rigid structures against which the actin filaments could exert a contractile force by the formation and breakage of cross-links. A Ca^{2+} -dependent ATPase is present in the isolated spindle in a concentration three times that in the cytoplasm.

The theories on the mechanisms involved in chromosome separation have arisen from observations of living or fixed, intact cells. Such studies are hampered by the inability to experiment directly with the mitotic apparatus, except as it exists in the complexity of its cellular environment. Exciting possibilities have been raised by recent successful attempts to isolate and experiment with the mitotic apparatus *in vitro*. Intact spindles from the eggs of the surf clam show normal sensitivity to temperature and colchicine. These isolated spindles can incorporate brain tubulin during their reassembly and such hybrid spindles are also cold labile, Ca^{2+} -sensitive and capable of considerable increase in overall length. Moreover, early anaphase spindles isolated from kangaroo rat cells will continue chromosome motion, in the *absence* of exogenous spindle subunits, when ATP is added. These results already suggest that while spindle growth requires microtubule polymerisation, anaphase movements do not.

REFERENCES

Beevers H., "Glyoxysomes of castor bean endosperm and their relation to gluconeogenesis", *Annals of the New York Academy of Sciences*, 1969.

Gibbs M., "Photorespiration, Warburg effect and glycolate", *Annals of the New York Academy of Sciences*, 1969.

Jackson W.A. & Volk R.J., "Photorespiration", *Annals of the New York Academy of Sciences*, 1970.

Tolbert N.E., "Microbodies-peroxisomes and glyoxysomes", *Annals of the New York Academy of Sciences*, 1971.

Zelitch I., *Photosynthesis, Photorespiration and Plant Productivity*, Academic Press, New York, 1971.

9

Structure and Development of Chloroplasts

Plastids are organelles which are bounded by double membranes and which occur, as far as is known, in all cells of eukaryotic green plants at some stage, usually becoming modified according to their function. In their undifferentiated form they may remain as proplastids, which are characteristic of epidermal and meristematic cells, for example. In the green parts of plants the proplastids normally develop into chloroplasts, which are the site of photosynthesis, while in starchstoring organs they form amyloplasts which produce the starch grains. However these two functions are not mutually exclusive as most chloroplasts will form starch under appropriate physiological conditions and the exposure of starch-storing organs to illumination results in the amyloplasts forming some thylakoids and chlorophyll. In certain plant parts, such as flowers, fruits and some leaves, the thylakoids of the chloroplasts become degraded, forming chromoplasts, which contain large amounts of carotenoids, the pigments responsible for 'autumn colouration' and the characteristic colours of certain flowers and fruits.

CHLOROPLAST STRUCTURE

Dimensions and Number

Beans possess chloroplasts typical of the sun leaves of

higher plants in that they are discoid or lens-shaped with a diameter of about 5μm and maximum thickness of about 1 μm, which is a vertical section through the disc. On the other hand observations of chloroplasts in living cells by phase contrast microscopy suggest that the maximum thickness of the chloroplast may be rather less. Under optimum conditions for growth, *Phaseolus vulgaris* averages 45 chloroplasts per palisade mesophyll cell and 32 per spongy mesophyll cell, giving about 8.3×10^8 chloroplasts per leaf and about 2.3×10^7 chloroplasts per cm^2 of leaf. This latter number is similar to that of mature spinach leaves although the number of chloroplasts per spinach mesophyll cell is more than ten times greater than for bean.

Mobile and Stationary Phases

From observations of chloroplasts in living cells by phase-contrast microscopy Wildman distinguished two components of the chloroplasts, a stationary phase which he equated to the thylakoid system and a mobile phase which he equated to the stroma. The mobile phase surrounds the grana and also penetrates the intergranal regions of the chloroplast. The mobile phase is always in some kind of motion though the intensity of the activity shows variability, even in the same chloroplast. In the reported investigations, observations were always made in cells which could be seen to be living by the presence of visible protoplasmic streaming. In such cells the mobile phase formed protuberances which sometimes broke away from the chloroplast and became indistinguishable from mitochondria, and also mitochondria-like bodies appeared to fuse with the mobile phase. We have found that in plants grown under low light intensities each chloroplast has one or more mitochondria embedded in deep pockets close to the chloroplast margin but that no break in the chloroplast envelope occurs and there is always at least a thin layer of cytoplasm between chloroplast and mitochondrion. It is possible that some of the observations of Wildman and colleagues may be explained by the development of the latter phenomenon.

Range of Chloroplast Structure

Quite early in the application of electron microscopy to plant structure the algae were found to show an interesting range of structural diversity of their chloroplasts. Subsequent studies on vascular plants established that a number of structural types were to be found in both cultivated and non-cultivated plants and furthermore that structural modification might be induced by mutation or by variation of the environmental conditions.

Dimorphic Chloroplasts of C_4 Plants

C_4 plants form oxalacetate, malate and aspartate as the primary products of their photosynthetic CO_2 -fixation in contrast to the more 'usual' C_3 plants whose primary fixation product is 3-phosphoglycerate. Laetsch lists the following families of flowering plants as containing C_4 species: Amaranthaceae, Aizoaceae, Chenopodiaceae, Compositae, Cyperaceae, Euphorbiaceae, Gramineae, Nyctaginaceae, Portulaceae and Zygophyllaceae. All of these families also contain C_3 plants and there is the case of the genus *Atriplex* where the C_4 *A. roseu* will hybridise with the C_3 *A. patula*. Apart from the difference in primary photosynthetic CO_2 -fixation products C_4 and C_3 plants show other substantial differences in their biochemistry, physiology, anatomy and fine structure. Maize and bean will be discussed here as typical examples of C_4 and C_3 plants respectively

In the maize leaf the chloroplasts are concentrated in two concentric sheaths of cells around each vascular bundle. The inner sheath of cells is described as the bundle sheath and consists of equal numbers of large barrel-shaped cells and smaller cells which can be divided into two sorts on the basis of their dimensions. Laetsch points out that the thick walls of bundle sheath cells adjoining mesophyll cells in C_4 grasses possess an electronopaque layer. The bundle sheath chloroplasts possess abundant thylakoids which do not associate into grana, and they normally contain starch grains.

The outer sheath of cells is known as the mesophyll sheath and it contains chloroplasts similar to those of C_3 plants in that they have grana. They are, however, different in that they do not normally contain starch grains. Not all C_4 plants show such structural dimorphism of their chloroplasts. However all chloroplasts of C_4 plants possess a system of tubules, the peripheral reticulum, which is associated with the inner membrane of the chloroplast envelope. There are reports of peripheral-reticulum-like membrane systems in chloroplasts of some cells of C_3 plants. Circumstantial evidence assembled by the latter author suggests that the function of the peripheral membrane in C_4 plants may be to facilitate transfer of metabolites between chloroplast and cytoplasm.

Effect of Environmental Conditions

The fact that plants grown in the complete absence of illumination form etioplasts while those grown under diurnal illumination form chloroplasts establishes that light is an essential requirement for chloroplast development. When Björkman *et al.*, compared plants of *Atriplex patula* which had been grown under three different irradiances: 20, 6.3 and 2 mW. cm^{-2} (in the waveband 400–700 nm), which are referred to as high, intermediate and low, they found that the high irradiance treatment yielded thicker leaves than the low, with more cells per leaf section and more chloroplasts per cell. The chloroplasts from plants grown under low irradiances were larger than those from high irradiance conditions and they contained more thylakoids and larger grana. The intermediate illumination gave intermediate results.

The low irradiance-grown *Atriplex* plants were compared with plants adapted for growth on the floor of the Queensland rainforest where the daily quantum flux was about one-twentieth of that provided by the low irradiance treatment. The chloroplasts of these plants, *Alocasia macrorrhiza, Cordyline rubra* and *Lomandra longifolia* were found to be dark green, unusually large and irregular in outline and to contain very well developed grana of enormous size. The grana were also irregularly arranged and apparently adapted for a maximum efficiency in light-trapping.

When maize plants were grown under very low irradiances (0.3 mW. cm^{-2}), which nevertheless provided about six times the daily quantum flux received by the rainforest plants, some chlorophyll and photosynthetic CO_2 -fixation developed even though the light compensation point was not exceeded and the leaves showed a net loss of CO_2 followed by senescence and premature death. Grana did not develop and both bundle sheath and mesophyll chloroplasts developed closely parallel arrangements of thylakoids which did not become appressed. Transfer of plants greened under very low irradiances to higher irradiances resulted in a partial recovery towards normal structure.

Research work in Belgium has shown that exposure of etiolated seedlings to electronic flashes of 1 millisecond in duration and given at 15 minute intervals brought about greening in which the resultant plastids were agranal with parallel unfused thylakoids. These flashed leaves showed the interesting property that, on transfer to continuous illumination, they initially did not show any photosynthetic oxygen evolution, but within 2 minutes of the commencement of illumination oxygen evolution was detected and a maximum rate was found after 6 minutes.

When etiolated leaves are transferred to continuous illumination without any pretreatment with light the onset of oxygen evolution tends to be considerably delayed. An alternative treatment which has produced almost completely agranal plastids is exposure of dark-grown seedlings to continuous far-red irradiation at wavelengths longer than 700 nm.

CHLOROPLAST MUTANTS

Many chloroplast mutants are known with defects in pigments or other components of structure or function. Some of the mutations are nuclear and show normal Mendelian inheritance whilst others show non-Mendelian inheritance and are considered to be mutations of the chloroplast DNA. Chloroplast mutations in algae can be maintained in heterotrophic culture and have been used for important research on chloroplast genetics, development and function. In contrast, although there have been numerous investigations of individual chloroplast mutations in higher plants, the only substantial research collection of higher plant material with mutant chloroplasts is the barley mutant collection of D. von Wettstein's group in Copenhagen. This collection has so far not been made generally available for studies on chloroplast development.

ISOLATION OF CHLOROPLASTS

Although Hill in 1937, obtained chloroplast suspensions from *Stellarsia media* which were capable of O_2 evolution under illumination, it was not until 1954 that Arnon *et al.*, reported the occurrence of photosynthetic phosphorylation in isolated chloroplasts. These latter chloroplast preparations were capable of rates of photophosphorylation and electron transport similar to those assumed to occur in intact leaves, but they showed very low rates of CO_2 fixation compared with the rates found in intact leaves (200 μmole CO_2 fixed/hour/mg chlorophyll). Little improvement in the rates of CO_2 fixation by chloroplast preparations was obtained until Walker devised procedares which initially gave spinach chloroplasts able to fix 24.3 µmole CO_2 /hour/mg chlorophyll. Subsequent modifications gave chloroplasts with improved rates of CO_2 fixation and in the author's laboratory the following adaptation of Walker's method has been used routinely. Washed spinach leaves are pre-illuminated for 30 minutes before use. 50 g of deribbed blades are rapidly cut up with a sharp knife to give pieces of about 2 cm^{-2} which are placed in a cooled perspex grinding vessel 6 cm × 5 cm × 25 cm.

The leaves are then covered with 200 ml of fresh grinding medium which has been partially frozen to a slushy consistency. The grinding medium consists of 400 mM sorbitol, 10 mM NaCl, 5 mM $MgCl_2$, 1 mM $McCl_2$, 2 mM EDTA, 2 mM isoascorbic acid, 0.4% (w/v) bovine serum albumin and 50 mM 2-(*N* -morpholino) ethanesulphonic acid (MES) adjusted to pH 6.5 (at room temperature). The leaves are ground for 3 seconds with a Polytron PT_{20} with a PT_{35} head at a speed setting of 3.5. The resultant slurry is squeezed through two layers of butter muslin and the filtrate poured through eight more layers of muslin. The final filtrate is centrifuged in an MSE Super Minor bench centrifuge with a precooled head. Rapid acceleration up to 4,000 × *g*, followed by braking by hand, enables the total centrifugation time to be less than 1 minute. The supernatant is discarded and the pellet gently resuspended with the aid of cotton wool and a glass rod in about 1 ml of solution identical to the grinding solution except that the isoascorbic acid is omitted and the MES is replaced by 50 mM *N* -2-hydroxyethylpiperzaine-*N′* -2-ethanesulphonic acid (HEPES) as buffer adjusted to pH 7.5 (at room temperature). The procedures are carried out at 0–4°C and the resultant chloroplast preparation is kept on ice.

The total preparation time from cutting the leaves should be about 4 minutes. Most of the chloroplasts (60–80%) in such a preparation should be complete chloroplasts with a morphologically and functionally intact envelope and a high rate of light-dependent CO_2 -fixation and O_2 -evolution, similar to that found *in vivo*. They are considered to possess a full complement of unimpaired photosynthetic reactions although most of the reactions cannot be measured directly as the inner membrane of the envelope is impermeable to most of the intermediates which might be added to test these reactions. S. G. Wildman has demonstrated to me, by phase contrast microscopy, that most of the chloroplasts in this sort of preparation have lost their starch grains and therefore presumably possess resealed membranes.

CHLOROPLAST DEVELOPMENT

Although the study of chloroplast development and the onset of photosynthesis in seedlings has attracted attention during essentially the whole of the twentieth century it is in recent years that most interest has been shown. There have been two main methods of conducting this study, of which the first is essentially concerned with chloroplast development in plants grown under natural environmental conditions. Since natural conditions are normally neither constant nor consistent such studies have often been conducted in controlled environmental chambers set to a standard day length and standard conditions for day and night. When it became evident that illumination was the critical environmental factor controlling chloroplast development it became fashionable to study chloroplast development by allowing the seedlings to grow for an initial period of several days in complete darkness prior to their transfer to continuous illumination.

Both approaches have their merits and it is clear to the present author that both are necessary to obtain an understanding of chloroplast development. Obviously seedling growth under natural environmental conditions shows the normal state of chloroplast development but it provides two main experimental difficulties. Firstly, under diurnal conditions there is a gradient of chloroplast development both within the plant and within the leaf, such that any experimental analysis is either very difficult or impossible. The second difficulty is that diurnal conditions make it difficult to distinguish the effects of illumination on chloroplast development.

If seedlings are grown in continuous darkness for a sufficiently long period, their development reaches a stationary phase in which all of the developing plastids in a leaf or a section of a leaf show the same stage of growth. On transfer to illumination the subsequent development then tends to occur in a synchronous manner, thus facilitating microscopic analysis and making biochemical analysis feasible. Such treatments may enable leaf chloroplast development to show synchrony like that obtainable in certain microbial cultures. It should also be pointed out that much progress in the study of photomorphogenesis has depended on illumination treatments of such darkgrown seedlings.

CHLOROPLASTS AND PHOTOSYNTHESIS

The formation of a functional thylakoid from the prolamellar body membranes of the etioplast during the greening process will also be considered at the molecular level.

Thylakoid Structure

Thin sectioning and heavy metal shadowing

When thin sections of fixed and stained leaf material are examined in the electron microscope, the chloroplast thylakoid membrane appears as a single electron dense line or as a tripartite structure (two dense lines with an intermediate light line) characteristic of the unit membrane. In high resolution electron micrographs of thin sections, fixed in permanganate, the thylakoid membranes give the appearance of containing globular subunits, 9 nm in diameter. It is possible, however, that these are an artefact since it is difficult to understand how several layers of overlapping 9 nm particles can be clearly resolved in a section thicker than 40 nm.

Some of the earliest electron micrographs of thylakoid membranes were obtained simply by drying isolated chloroplast fragments of *Spirogyra* and *Mougeotia* on an electron microscope grid and shadowing the preparations with chromium. The membranes showed a granular or particulate structure. Similar observations were made subsequently with spinach thylakoid membranes. The particles usually existed in a random array but sometimes they were seen in highly ordered arrays. Their dimensions were 15 × 18 × 10 nm and they appeared to contain subunits. Park suggested that the particle, which he termed quantasome, might be the morphological expression of the so-called photosynthetic unit, defined as the minimum number of chlorophyll molecules needed to fix one molecule of carbon dioxide per intense flash of light. The quantasomes were seen on the inner surface of the thylakoid membrane, suggesting that there was some distortion of the thylakoid membrane during drying.

Freeze-Etching

The freeze-etch technique is a different approach to the study of the substructure of the thylakoid membrane, because chemical fixatives are not used. Leaf material or isolated chloroplasts are frozen and fractured. The fracture plane in the frozen sample occurs along hydrophobic regions within the thylakoid membrane exposing complementary faces. Platinum-carbon replicas of the fracture planes are made and examined in the electron microscope.

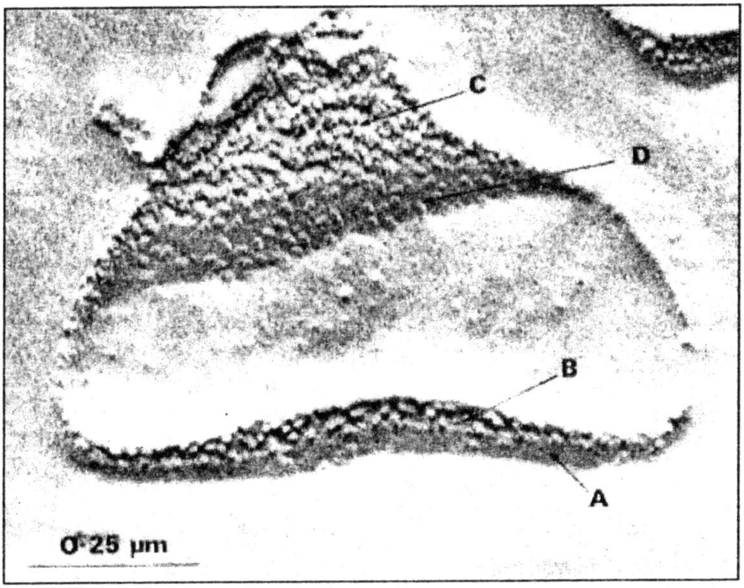

Figure .1a. Freeze-fracture electron micrograph of a spinach grana thylakoid, showing the large and small particle fracture faces, B and C, and the surfaces, A' and D.

Two major sizes of particles are seen when mature plant chloroplasts are freeze-fractured; large particles of average diameter, 17.5 nm and small particles of average diameter, 11 nm. (Fig. 1a) The small and large particles occur on different fracture faces. (Fig. 1b) Fracture face C, which is towards the stroma of the chloroplast contains tightly packed small particles, and the complementary fracture face B contains the large particles at about half the density per μm^2. The large particles are only observed within grana stacks, where the adjacent thylakoid membranes are in close contact, while the small particles are observed in grana and stroma thylakoids. In the unstacked stroma thylakoids the complementary fracture face (designated B_u) is relatively smooth and shows only a few small particles.

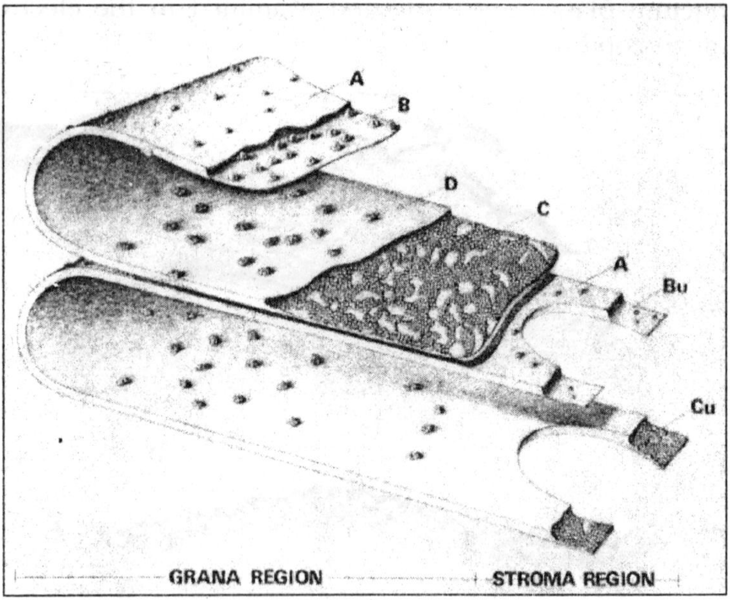

Figure 1b . Model of the thylakoid membrane, showing grana and stroma regions.

The exterior thylakoid surface, A′, and the interior surface, D, are observable by the deep etch technique, in which the ice covering the surfaces is sublimed off before the replica is made. In well-washed thylakoids the A′ surface shows little surface relief, but is covered with proteins in unwashed lamellae. The interior surface is covered with particles about 18.5 × 15.5 nm in size which show slight surface relief. These particles on the D surface contain 3 or 4 subunits and they may correspond to the quantasomes as seen by metal shadowing.

The freeze-etch data suggest that the particles within the thylakoid membrane are arranged somewhat asymmetrically. The larger particles are located towards the inner surface and partly protrude into the intrathylakoid space, and the smaller particles are near the outer surface. The model of the thylakoid membrane shown in Fig.2

appears to be consistent with the freeze-etching results and also with X-ray diffraction studies of thylakoid membranes.

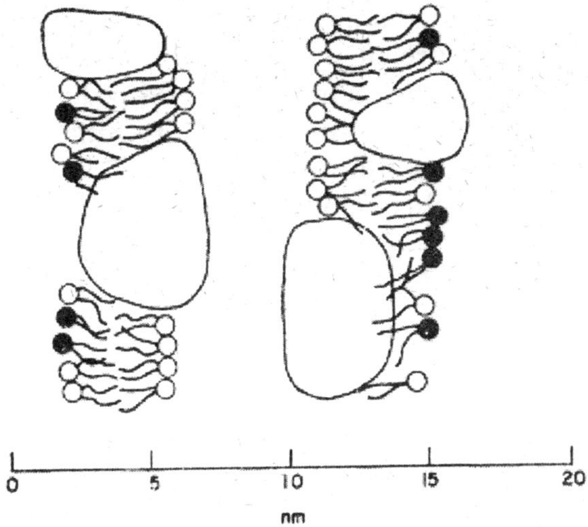

Figure 2. Structural model of the thylakoid membrane showing an asymmetric distribution of large and small particles.

The membrane is viewed as a lipid bilayer in which the particles are embedded, in accordance with the fluid mosaic model for biological membranes.

Negative Staining

Two sizes of particles are observed on the surface of unwashed thylakoids by negative staining with phosphotungstic acid. The larger particles (11 nm in diameter) are removed by water washing the thylakoids and correspond to the CO_2 -fixing enzyme, ribulose bisphosphate carboxylase. The smaller particles (9 nm) are removed by washing with the chelating agent, ethylenediamine tetraacetate, and are identical with the coupling factor (CF_1) which catalyses the terminal steps in the energy coupling process of ATP formation. These surface particles are not related to the membrane particles seen by freeze-etching.

Thylakoid Composition

Lipids

The thylakoid membrane consists of 50% protein and 50% lipid. The feature of its lipid composition is the high percentage of the neutral galactolipids, digalactosyl diglyceride (27%) and monogalactosyl diglyceride (13%), compared with the phospholipids (9%). Table 1) Chlorophyll a plus chlorophyll *b* account for 21% of the lipids.

Table 1. Lipid composition of spinach thylakoid membranes

	% Weight of total lipid	*Number of lipid moles per 400 chlorophyll moles*
Monogalactosyl diglyceride	26.8	602
Digalactosyl diglyceride	13.4	198
Sulpholipid	4.1	42
Phospholipid	9.1	201
Chlorophyll *a*	15.0	290
Chlorophyll *b*	5.8	110
Carotenoids	2.8	82
Quinone compounds	3.3	80
Sterols	2.2	
Unidentified lipids	17.0	

The thylakoid membrane also contains an anionic sulpholipid. The other unique feature of the thylakoid lipids is their high content of the polyunsaturated fatty acid, linolenic acid (18:3).

Chlorophyll-Proteins

The proteins of the thylakoid membrane include the carriers of the photosynthetic electron transport chain, but these account for only a relatively small percentage. When a thylakoid preparation, from which the coupling factor, lipids and pigments have been removed, is solubilised by the anionic detergent, sodium dodecyl sulphate (SDS), and the extract examined by SDS-polyacrylamide gel electrophoresis, at least 20 polypeptides are obtained. These range in size from 10,000 to 100,000 daltons. A major fraction of the proteins appear in the region of 25,000 daltons. If the thylakoids are extracted with SDS without prior removal of pigments, two chlorophyllprotein complexes are separated by SDS polyacrylamide gel electrophoresis. Complex I which is derived from photosystem 1 has a chlorophyll *a* /chlorophyll *b* ratio of 12:1 and accounts for about 20% of the thylakoid membrane protein. It has an apparent molecular weight of 110,000 daltons and contains 14 moles chlorophyll *a* per mole of protein. Complex 11, which accounts for 40–50% of the thylakoid membrane proteins has an apparent molecular weight of 32,000 daltons and contains one mole chlorophyll *a* and one mole of chlorophyll *b* per mole of protein.

REFERENCES

Arntzen C.J. & Briantais J.M., "Chloroplast structure and function", In *Bioenergetics of Photosynthesis,* (ed. Govindjee). Academic Press, New York, 1975.

Calvin M. & Bassham J.A., *The Photosynthesis of Carbon Compounds,* Benjamin, New York, 1962.

Heber U., "Metabolite exchange between chloroplasts and cytoplasm", *Annual Review of Plant Biology,* 1974.

Kirk J.T.O. & Tilney-Basset R.A.E., *The Plastids,* Freeman, London and San Francisco, 1967.

Park R.B. & Sane P.V., "Distribution of function and structure in chloropalst lamellae", *Annual Review of Plant physiology,* 1971.

Trebst A., "Energy conservation in photosynthetic electron transport of chloroplasts", *Annual Review of Plant physiology,* 1974.

10

Functions of Plant Cell Walls

Plant cell walls establish a home, and indeed a city, for plant protoplasts. They serve many specialised functions in plant tissues, and form the skin, the skeleton, and the circulatory system of plants.

There are many variations in the form and substance of plant cell walls. The walls may be plastic or they may be rigid, permeable or impermeable, impregnated with plastics or coated with slime, cemented in layers to form fibres or dissolved in spots to form pores. These variations are of vital importance to the proper biological functioning of plant cells and organs, and thus the structure of the cell wall is often our best indication of the nature of the protoplast which dwells inside.

MOLECULAR STRUCTURE OF PLANT CELL WALLS

Polysaccharides are the principal components of all plant cell walls. The polysaccharides of the cell wall are made up of sugars which are linked to each other by glycosidic bonds to form the polymer chains. Each polysaccharide contains particular kinds of sugars which are joined to each other in characteristic patterns of linkage position and sequence. It is now known that the secondary, tertiary and quaternary structures of cell wall polysaccharides are determined by the structures of the component sugars and the linkages between them, just as the three dimensional structure of a protein is determined by the sequence and structures of its component amino acids.

The various polysaccharide chains of the plant cell wall are connected to each other in specific ways, and they form an integrated network. The properties of this network depend not only on the amounts, characteristic properties and orientations of the individual polysaccharides, but also on the nature and frequency of the interconnecting linkages between them.

The three types of polysaccharide normally found in plant cell walls (cellulose, hemicelluloses, and pectic polysaccharides), and the structural protein of primary walls, are described briefly below.

Cellulose

Cellulose occurs as a crystalline, fibrillar aggregate of *b* -1,4-linked glucan chains. Cellulose fibrils give plant cell walls most of their enormous strength, much as glass fibres embedded in an epoxy resin give strength to a fibreglass composite.

The basic structure of the *b* -1,4-linked glucan chains of cellulose is illustrated in Fig. 1 by conformational line drawings and in Fig. 2 by molecular models. Residues of *b* -D -glucose are glycosidically linked to each other, *from* carbon 1 of one residue *to* carbon 4 of the adjacent residue. The upside-down inversion of every second residue in the chain minimises contact between atoms of adjacent residues. Close inspection of the models in Fig. 2 shows that the –OH groups at carbon 3 are in very close proximity to the ring oxygens (O_5) of adjacent residues. Hydrogen bonds between O_3 and O'_5 help to stabilise the flat, straight, ribbon-like structure of *b* -1,4-linked glucan chains.

The flat, ribbon-like structure allows the chains to fit closely together, one on top of the other, over their entire lengths. These *interchain* associations are stabilised by hydrogen bonds between O_6 of a glucose residue in one chain and the oxygen of the glycosidic bond (O_1) between glucose residues in an adjacent chain. Since the glucan chains of cellulose are very long (8,000 to 15,000 residues), the number of hydrogen bonds between adjacent chains is very large. The resultant crystal is extremely stable and so tightly packed that there is no room for water molecules in the crystal structure.

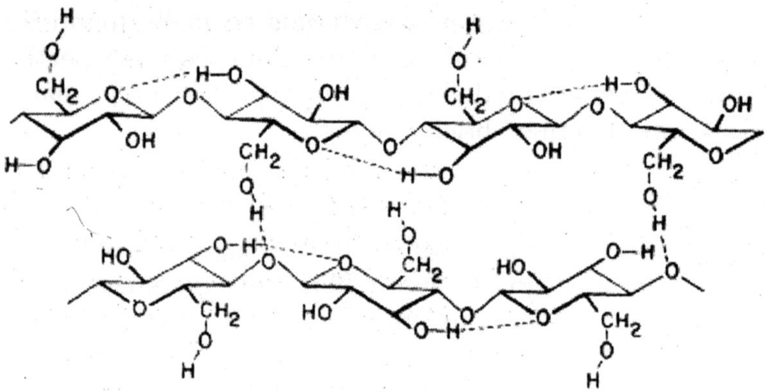

Figure 1. b -1,4-linked glucan chains of cellulose.

Although there is some controversy as to whether native cellulose fibrils are 3.5 nm or 10 nm in diameter, it is clear that the glucan chains of cellulose aggregate to form stiff crystalline rods of very considerable length and mechanical strength. The molecular structure of *b* -1 ,4-linked glucose thus neatly determines the secondary and tertiary structures of cellulosic glucan chains, and establishes the quaternary, interchain associations—although not the dimensions—of the microfibrils. The stiff, crystalline rods of cellulose are clearly well suited to their biological function in the plant cell wall.

Hemicelluloses

Xylans, arabinoxylans, galactomannans, glucomannans

and xyloglucans are common types of hemicelluloses. The basic repeating sequence of a hemicellulosic xyloglucan molecule is shown in Fig. 2, adjacent to a cellulosic glucan chain.

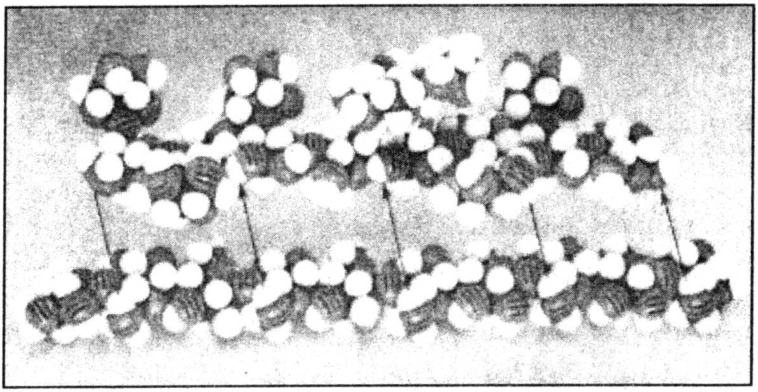

Figure 2. Hemicellulosic xyloglucan associated with cellulose.

Although different hemicelluloses have different component sugars, all have two structural features in common which bear importantly on their biological function. (1) All hemicelluloses have straight, flat *b* -1,4-linked backbones. Any side chains attached to the backbone are short—usually just one sugar long—and stick out to the sides of the backbone (Fig. 2). (2) All hemicelluloses have some structural feature which prevents the chains from extended self-aggregation of the type which exists between the *b* -1,4-linked glucan chains of cellulose.

Xyloglucans, for example, have a *b* -1,4-linked glucan backbone (just as cellulose), but most of the glucose $-CH_2$ OH groups (C_6) are substituted with xylose side chains (Fig. 2) and are thus not available for the formation of interchain hydrogen bonds. Similarly, since xylose is a pentose (*i.e.* a five carbon sugar), the *b* -1,4-linked xylose backbones of xylans and arabinoxylans have no–CH_2 OH groups available for interchain hydrogen bonding. The glucomannans have *b* -1,4-linked backbones containing both glucose and mannose. The $-CH_2$ OH groups of these sugars are unsubstituted, but the axial conformation of the –OH groups at carbon 2 of the mannose residues prevent close interchain associations wherever mannose residues occur in the backbone.

Although hemicelluloses cannot self-aggregate to form long, close-packed crystalline fibrils in the manner of cellulose, the chains of hemicellulosic polysaccharides can form important hydrogen bonded associations with each other in the cell wall, particularly between regions of the chains which have few side branches or axial hydroxyl groups. Even more importantly, hemicelluloses can co-crystallise with cellulosic glucan chains at the surface of the cellulose microfibrils. The cocrystallisation probably involves the formation of hydrogen bonds *from* the $-CH_2$ OH groups present in cellulose chains *to* the glycosidic oxygens in the adjacent hemicellulose chains (Fig. 2). This association would form a tightly bound monolayer of the hemicellulose on the surface of the cellulose microfibril, and would function as part of the 'glue' which holds the microfibrils together in the cell wall.

Pectic Polysaccharides and Structural Protein

Rhamnogalacturonans

The rhamnoglacturonans are long polymers of *a* -1,4-linked galacturonic acid interspersed with a few residues of 1,2-linked rhamnose. There is some evidence that the rhamnosyl residues may occur at definite positions in the galacturonan chain, giving a subunit structure to the rhamnogalacturonan polymer (Fig. 3).

The diaxial conformation of the *a* -1,4 glycosidic linkages between galacturonic acid residues causes the orientation of adjacent rings to be twisted. Thus, the galacturonan polymer forms a tight, stiff, rod-like helix with three residues per turn (Fig. 3). The insertion of 1,2-linked rhamnosyl residues in the galacturonan chain creates 'kinks' or right angle bends.

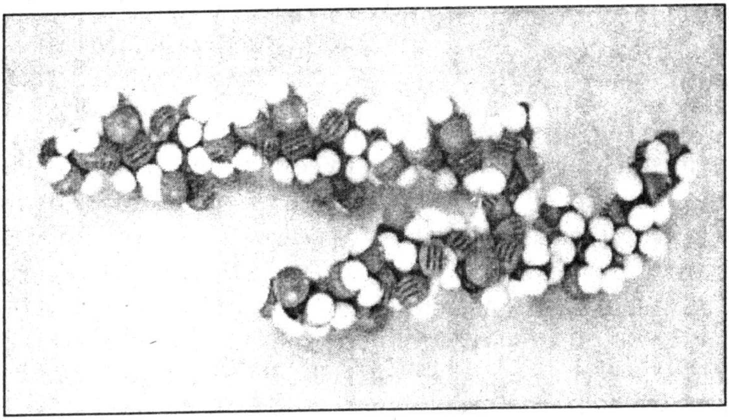

Figure 3. Pectic rhamnogalacturonan.

Divalent cations, particularly calcium, form complexes with the carboxyl and hydroxyl groups of galacturonic acid residues in the polymer. Complex formation of this type occurs primarily between adjacent residues in a galacturonan polymer, but could serve to create ionic ligand bridges between adjacent galacturonan chains.

Arabinogalactans

Two distinct types of arabinogalactans are known to occur in plant cell walls. The first type has a *b* -1,4-linked galactan backbone with highly branched arabinose side chains. The second type has a *b* -1,3-linked galactan backbone with many short side chains containing galactose and arabinose.

Structural Protein

Primary cells walls contain a structural protein component. While the structural protein of these walls has not yet been isolated as an intact molecule, the analysis of peptide fragments from the primary walls of dicotyledonous plants has revealed several interesting characteristics. This structural protein contains over 25% hydroxyproline, which is an unusual amino acid known to break the continuity of *a* -helical structures. In animals, hydroxyproline occurs almost exclusively in the proteins of connective tissue (collagen and gelatin).

Several tryptic peptides of the structural protein have been isolated, and all appear to contain the sequence Ser-Hyp_4. Arabinose tetrasaccharides are glycosidically linked to virtually all of the hydroxyproline residues in the protein. In addition, many of the serine residues are glycosylated with galactose. As a result of this extensive glycosylation, the structural protein is very resistant to degradation by proteases, and is likely to have an extended, rod-like shape.

INTERCONNECTIONS BETWEEN CELL WALL COMPONENTS

The polysaccharide components of the primary cell walls of dicotyledonous plants are specifically attached to one another to form an interconnected network. The *b* -1,4-linked glucan chains of cellulose are aggregated to form crystalline microfibrils. The reducing ends of the xyloglucan chains are covalently (glycosidically) attached to *b* -1,4-linked arabinogalactan. The arabinogalactan chains, in turn, are glycosidically linked to the backbone of the rhamnogalacturonan, probably to the 4 positions of the rhamnosyl residues.

The 1,4 arabinogalactan is thus a side chain of the rhamnogalacturonan, and forms an interconnecting bridge between the xyloglucan and the rhamnogalacturonan. The other type of arabinogalactan (*b* -1,3-linked) also seems to serve as an interconnecting bridge, glycosidically linking the reducing ends of rhamnogalacturonan chains to amino acid residues in the structural protein of the cell wall.

UNIVERSALITY OF PLANT CELL WALL STRUCTURES

Studies of cell walls isolated from suspension-cultured plant cells have demonstrated quite clearly that dicotyledonous plants from taxonomically diverse species have very similar primary walls. Similar studies on the primary walls of aspen cambial tissue and potato tuber tissue have shown that these walls are quite similar to the walls of suspension-cultured cells. Thus, all dicotyledons probably have the same basic primary cell wall.

The primary cell walls of monocotyledons have a structure which is somewhat different from that of dicotyledons, although the walls from various monocotyledon species appear to be quite similar to each other. The hemicellulose of the primary monocotyledon walls is an arabinoxylan instead of a xyloglucan, and the structural protein appears to contain very little hydroxyproline.

Most cells, after they stop active growth, synthesize 'secondary' cell wall material which is quite different from the primary wall. Secondary cell walls consist almost exclusively of large amounts of cellulose and one (or more) of the hemicellulosic polysaccharides. Lignification is very common for cells with secondary walls.

Within broad taxonomic groups, related plants appear to have very similar cell walls. However, it should be recognised that relatively few types of cell walls from relatively few plant species have been carefully analysed. The diversity of specialised cell wall functions is likely to be reflected in a diversity of cell wall structures.

Cell Wall Plastics

Lignin

Lignin is a biological plastic formed in plant cell walls by the enzymic dehydrogenation of coumaryl, coniferyl and synapyl alcohols (Fig. 4) followed by a free radical polymerisation. Since the polymerisation is not enzymically controlled, and the monomeric free radicals can react with each other in a variety of ways, lignin does not have a unique structure. However, the lignin from a particular plant species or tissue does usually have a characteristic monomer composition. Some covalent linkages are formed between lignin and the polysaccharides of the cell wall during lignin biosynthesis.

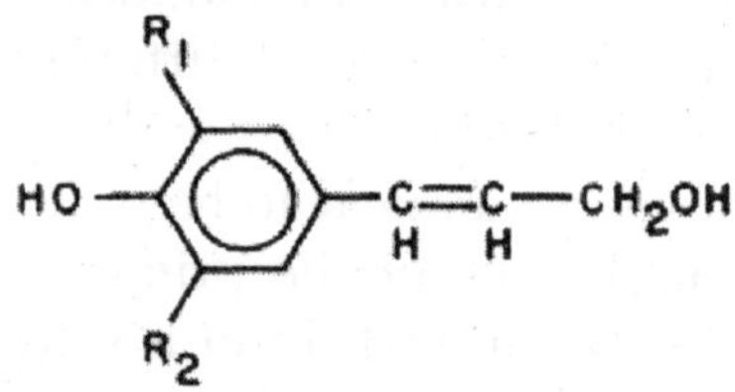

Figure 4. Lignin precursors

Lignin formation is initiated very soon after secondary wall synthesis begins, and proceeds from the region of the middle lamella inward towards the plasmalemma. Thus, both the primary and secondary walls become fully impregnated with a rigid, hydrophobic plastic which is covalently linked to the polysaccharide matrix. The resultant structure is extremely strong and resistant to degradation.

Cutin

Cutin is a biological plastic which coats the cell walls on the outer surface of plant epidermal cells. Although little is known about the detailed structure of cutin, it is clear that the principal monomeric components of this material are mono-, di-, and trihydroxyfatty acids (C_{16} –C_{18}). These hydroxyfatty acids are linked to each other mainly through ester bonds, although the presence of both ether and peroxide bonds have been reported. The formation of ester linkages between monomeric hydroxyfatty acids and preformed cutin appears to involve the enzymic transacylation of the hydroxyfatty acids from a Coenzyme A intermediate to free hydroxyl groups in cutin.

The film of cutin polymer is impregnated—and frequently coated—with a complex mixture of waxes. The cutin-wax structure (cuticle) merges gradually with the normal polysaccharide components of the epidermal cell wall. There is evidence for the existence of hydrophylic channels (ectodesmata) in the hydrophobic cuticle. These channels may occur in morphologically distinct patterns on the epidermal surface.

The cuticular waxes and the cutin polymer form a tough hydrophobic skin over the surface of the plant which is important in minimising water loss and preventing mechanical injury or invasion by pathogens.

BIOSYNTHESIS OF PLANT CELL WALLS

It is generally believed that cell wall polysaccharides are synthesized from the appropriate sugar nucleotides (e.g. UDP-glucose) by specific enzymes or enzyme complexes. Many—though not necessarily all—cell wall polysaccharides are synthesized in the golgi bodies (dictyosomes) and transported to the wall in golgi-derived vesicles which are able to fuse with the plasmalemma. The cell wall polysaccharides are then presumably interconnected while in the wall.

There are many reports of isolated enzymes or particulate complexes which incorporate sugars from the sugar nucleotides into polysaccharide material. However, the amount of polysaccharide material formed by these *in vitro* enzyme systems is often a miniscule fraction of the amount formed *in vivo*. Thus the relationship between these isolated enzymes and cell wall biosynthesis is uncertain.

Isolated enzyme preparations have been reported to synthesize alkaliinsoluble *b* -1,4 glucan ('cellulose') at rates comparable to the rate of *in vivo* cellulose synthesis. However, these enzymes have not as yet been shown to form the microfibrils characteristic of cellulose. There is some evidence that cellulose microfibrils are synthesized at the plasmalemma rather than in the golgi, perhaps by plasmalemmafused vesicles which contain rows of membrane-bound, microfibril-synthesizing particles.

CELL WALL ULTRASTRUCTURE

Cellulose Microfibrils

One of the most important and intriguing aspects of plant cell wall ultrastructure is the orientation of the cellulose microfibrils within the wall. Cellulose microfibrils are deposited (or possibly synthesized) at the inner surface of the wall, adjacent to the plasmalemma. They have a particular orientation when deposited. On the other side of the plasmalemma there is a thin layer of microtubules. The orientation of the microtubules in this layer exactly parallels the orientation of the cellulose microfibrils being deposited on the opposite side of the plasmalemma. Disruption of the microtubules (e.g. by colchicine) affects the orientation of the microfibrils.

In the primary walls of elongating cells, the most recently deposited cellulose microfibrils (those closest to the plasmalemma) lie parallel to the plasmalemma and are oriented perperndicular to the long (growth) axis of the cell, like the hoops which hold a barrel together. Cellulose microfibrils deposited at earlier times (and thus further from the plasmalemma) are still parallel to the plasmalemma, but are oriented at decreasing angles (i.e. more nearly parallel) to the growth axis of the cell. It is as though the process of cell wall elongation pulls the ends of the microfibrils towards the ends of the cell. In the end walls of a cell, or in the newly forming cell wall created by cell division (i.e. the cell plate), the cellulose microfibrils again lie parallel to the plasmalemma, but are randomly crossed and not parallel to each other.

The cellulose microfibrils in secondary walls are usually deposited in discrete, concentric layers. The microfibrils in a given layer are all parallel to each other, but are oriented at a considerable angle to both the microfibrils in adjacent layers and to the long axis of the cell. This cross-hatching of microfibrils in adjacent layers of secondary walls adds considerably to the overall strength of the wall, and is the same principle used in the manufacture of fibreglass-epoxy fishing rods, where great strength for weight is essential.

Secondary wall material may also be deposited in very localised regions to form highly specialised structures such as the rings or spirals of secondary wall found in xylem cells. In such cases the microtubules on the inside of the plasmalemma have the same localisation and orientation as the cellulose microfibrils being deposited on the opposite side of the plasmalemma.

Specialised Structures

Many of the modifications of cell wall structure which occur are related to communication or transport between cells. In cells with primary walls, intercellular communication is facilitated by plasmadesmata. Plasmadesmata are small pores of approximately 40 nm diameter which extend through the wall between two adjacent cells. The plasmalemma of the cell(s) also extends through these pores, so that the cytoplasm of one cell is continuous with that of adjacent cells. There appears to be a core of electron dense material in the center of the plasmadesmatal tubes. This core may be a specific (though as yet speculative) structure capable of regulating what can and cannot be passed between cells. Regardless of the nature of the core, it is clear that the plasmadesmata are an aspect of plant biology which has no obvious counterpart in animal tissues. Each cell may have thousands of plasmadesmata, either randomly scattered as individual pores or grouped in distinct fields where the primary wall is often thinner than usual.

In cells with secondary walls, and particularly in tracheids, intercellular communication and transport is facilitated by pits. In the full, bordered type of pit, the primary wall and middle lamella are largely dissolved. The hole thus formed is surrounded and partly overgrown by depositions of secondary wall material which create a raised, overhanging border. The primary wall and middle lamella in the pit is replaced by a relatively impermeable disc or torus of thickened wall material. The disc is suspended by an easily permeable, radial network of cellulose microfibrils. The diameter of the disc is slightly larger than the aperture of the overhanging pit border. Thus the disc can act as a valve to close the pit when pressed against the border by a large pressure differential. Gas bubbles, which would break the flow of water through the tracheids, can be sealed off by the action of such valves.

Where the tracheids are adjacent to ray cells, the pits are differentiated only on the tracheid side, giving half-bordered pits. The walls of the ray cells in the pit region appear to remain intact or only slightly modified.

A further important example of a cell wall modification which facilitates intercellular communication or transport is the sieve plate of primary phloem. The phloem cells have primary cell walls with many plasmadesmata. Some of the plasmadesmata of the crosswalls between adjacent phloem cells enlarge considerably and become lined with callose (a *b* -1,3 glucan) on the *wall* side of the plasmalemma. The pores of the sieve plate thus formed become filled with a fibrous protein.

Many other specialised cell wall structures or modifications could be described. Several examples may be mentioned just to indicate the range of possibilities: the walls of pollen cells, which are so indestructible that they are used by archeologists to characterise the flora of dwelling sites many thousands of years old; the walls of bark cells, made largely of suberin, a wall material similar to cutin that can form a protective coating over wounds so as to prevent water loss or pathogen invasion; and the walls between adjacent endodermal cells, which have very dense Casparian strips which prevent the movement of ions through the walls and into the xylem stream.

HORMONAL CONTROL OF CELL WALL BIOSYNTHESIS AND DIFFERENTIATION

We know that cell wall biosynthesis and the modification of cell walls are important aspects of cellular differentiation, and that cellular differentiation can be controlled—or at least affected—by plant hormones. However, the mechanisms by which particular plant hormones affect the form, substance and synthesis of particular cell walls are almost wholly unknown.

Significant progress has been made recently in elucidating the role of auxins in cell wall elongation. Indoleacetic acid and several similar compounds cause a marked increase in the rate of coleoptile cell elongation. Coleoptiles normally receive auxin from the apical region of the shoot. However, excised coleoptiles can respond to auxin exogenously supplied in solution, elongating at a rate of 10–30% per hour. The mechanism of auxin-stimulated elongation of coleoptile cells has been proposed to involve the activation of a hydrogen ion 'pump' in the plasmalemma. The hormonal activation of this pump results in a lower pH in the cell walls which, in turn, appears to activate enzymes in the cell wall which are capable of selectively 'loosening' the polysaccharide network so that the walls can elongate more rapidly. Auxin must also stimulate cell wall biosynthesis since the walls retain a constant thickness while more than doubling their length during prolonged auxin treatment.

A further important aspect of hormonal effects on plant cell wall biosynthesis and differentiation has been revealed by studies on the changes in microtubule orientation caused by exogenously supplied hormones. In bean epicotyl segments, kinetin inhibits elongation and promotes a thickening or lateral expansion of the cells. Gibberellins, on the other hand, promote elongation and inhibit lateral expansion. The microtubules adjacent to the plasmalemma in cells of epicotyl sections supplied with kinetin and auxin are found to be oriented parallel to the cell axis. However, in sections supplied with gibberellin and auxin the microtubules are oriented transverse to the cell axis.

The microtubules are randomly oriented in sections supplied with auxin alone. Thus, kinetin and gibberellins (but not auxins) appear to be able to control the direction of cell growth by somehow determining the orientation of the microtubules. The orientation of the microtubules, in turn, determines the orientation of the cellulose microfibrils being deposited in the cell wall—which determines the direction in which the walls can most easily expand.

INTERACTION OF PLANT CELL WALL IN OTHER ORGANISMS

Plants are beset by many pests and pathogens and helped by a variety of symbionts. The walls of epidermal cells may be specialised in a number of ways to form a protective skin over the entire plant. Many cell walls throughout the plant can become resistant to degradation by pathogens by means of lignification.

Quite apart from such modifications, however, the structure of the cell wall can itself present a formidable barrier to pathogens. In order to penetrate the cell wall and utilise its component sugars, pathogenic fungi and bacteria have evolved sophisticated, inducible batteries of enzymes which can hydrolyse components of the plant cell wall. The pathogen appears to have dissolved a hole through the cuticle and to have started degrading the cell wall.

Plants, in turn, have evolved countermeasures to the hydrolytic enzymes of the pathogens. In primary walls, the molecular architecture is such that only the pectic polysaccharides are accessible to hydrolytic enzymes. It is probably for this reason that the first enzymes to be secreted by an invading pathogen are the pectin-hydrolysing enzymes. This is also likely to be the reason why the walls of many plants contain proteins which specifically inhibit the pectin-degrading enzymes of pathogenic micro-organisms. In other plants the pectic polysaccharides are heavily acetylated, and thus resistant to enzymic attack.

Plants have evolved mechanisms for counterattack as well as defence. Inducible enzymes are present in the cell walls of several plants which hydrolyse the wall polysaccharides of invading fungal pathogens. The pathogens react by secreting proteins which specifically inhibit the attacking plant enzymes. The cell walls of a variety of other plants contain glycoside-hydrolysing enzymes which can release hydrogen cyanide from cyanogenic glycosides. The release of hydrogen cyanide by the plant occurs in response to attack by a pathogen. The pathogen (sometimes) avoids cyanide poisoning by an inducible enzyme which converts the cyanide to harmless formamide.

From these and other examples it is clear that the cell wall is a most important battleground in the contest between plants and their pathogens. The plant cell wall is not just a strong but passive barrier to invasion. It is impregnated with a host of molecules which can recognise a pathogen, modify the defences, or mount a counterattack.

REFERENCES

Albersheim P., "Substructure and function of the cell wall", In *Plant Biochemistry* (Ed J.E. Varner), Academic Press, New York, 1965.

Aspinall G.O., "Carbohydrate polymers of plant cell walls", In *Biogenesis of Plant Cell Wall Polysaccharides* (Ed. F. Loewus) pp. 95-115, Academic Press, New York, 1973.

Bateman D.F. & Basham H.G., "Degradation of plant cell walls and membranes by microbial enzymes", In *Physiological Plant Pathology* Vol. I (Ed. P.H. Williams & R. Heitifus) Springer-Verlag, Berlin, 1975.

Keegstra K., Talmadge K.W., Bauer W.D. & Albersheim P., "The structure of plant cell walls III. A model of the walls of suspension-cultured sycamore cells based on the interconnections of the macromolecular components", *Plant Physiology*, 1973.

Lamport D.T.A., *Cell wall metabolism*, Annual Review of Plant Physiology, 1970.

11

Membrane Structure and Transport in Plants

All compartments of the cell, and its external surface, are bounded by membranes. It is clear, therefore, that any complete understanding of control mechanisms in metabolism or development must include a precise knowledge of the structure and composition of membranes and of the mechanisms whereby materials move through them.

CHEMICAL COMPOSITION OF MEMBRANES

Biological membranes are composed primarily of two main classes of compounds, lipids and proteins, which interact in several ways with water to bring about a characteristic trilaminar structure. As a very broad generalisation it might be said that lipids make up the supporting matrix of the membrane while the proteins thus supported determine its specific properties. This is, of course, merely a convenient simplification as it is becoming plain that some of the characteristic properties of a membrane, particularly those which determine its effectiveness as a barrier to the diffusion of water and electrolytes, depend on the nature of the membrane lipids. Conversely some proteins have a structural role. For the present purposes, however, the classes of compounds will be considered separately and a synthesis attempted later on.

Lipids

Of the lipids present in plant cells the various phospholipids, glycolipids and sterols are of the greatest significance in membrane construction. The relative abundance of the components can be quite variable depending on the part of the plant analysed, the species and prevailing environmental conditions.

Phospholipids

The phospholipid molecule can be separated into a charged or polar 'head' region and an uncharged or non-polar 'tail'. Such a molecule is described as amphipathic, and as we shall see later on, this property is of crucial importance in determining membrane structure. Phospholipids are generally thought to be restricted to membranes but the extremely rapid rate at which membranes can be taken apart and re-assembled, as in cell plasmolysis and de-plasmolysis, makes it probable that there are stores of phospholipid within the cell.

Phospholipids are readily extracted from macerated plant tissues by a mixture of chloroform and methanol (2:1) and can be separated by thin layer chromatography using a variety of solvent systems.

The commonest phospholipids in plant membranes are derivatives of phosphatidic acid; thus, lecithin is the choline ester of phosphatidic acid. Phosphatidic acid (PA) itself is generally said to occur only in minute quantities in membranes or not at all, indeed, its presence in an extract is often taken as an indication of the activity of phospholipase D. There is a report, however, in which phosphatidic acid is said to be one of the major constituents of the plasmalemma of oat (*Avena sativa*) root. Unfortunately, detailed analyses of the plasmalemma from other plants are not available for comparison. In passing it might be noted that a great deal remains to be done, firstly in preparing pure sub-cellular fractions of the plasmalemma and of other membranes from plants and subsequently in determining their lipid composition.

Table 1 presents some of the available information on the distribution of the different types of phospholipid. The information on the composition of the membranes of mitochondria and chloroplasts is the most detailed and reliable since these organelles can be separated with relative ease and high purity during cell fractionation.

Table 1. Relative abundance of phospholipids in tissues and organelles.

	PA	*PI*	*PC*	*PE*	*PG*	*DPG*	*Other*
	(% of total phospholipid)						
Whole leaves							
Sugar Beet	ND	11	47	23	19	ND	—
Maize	ND	8	30	16	31	15	—
Bean	5	7	42	20	26	ND	—
Chloroplasts							
Sugar Beet	10	9	35	13	33	ND	—
Tobacco	ND	4	23	16	58	ND	—
Spinach (lamellae)	trace	10	20	13	57	ND	—
Mitochondria							
Cauliflower (floret)	ND	7	44	34	3	12	—
Potato (tuber)	ND	13	44	26	ND	17	—
Microsomes							
Cauliflower (floret)	ND	6	50	35	8	1	—
Potato (tuber)	ND	19	44	18	ND	19	—
Plasmalemma							
Oat (root)	14	4	18	11	4	ND	11

The predominant phospholipid in a given membrane may be characteristic, e.g. phosphatidylglycerol (PG) is a major component of chloroplast membranes while it is only a minor component of the inner mitochondrial membrane where diphosphatidylglycerol (DPG) is predominant. In general extracts of shoots and roots neither of these phospholipids is as abundant as lecithin (PC) or phosphatidyl ethanolamine (PE). A small quantity of phosphatidyl inositol (PI), usually less than 10% of the total phospholipid, is found in all membranes.

Organisms which live in warm conditions and warm blooded animals are generally found to have phospholipids with an abundance of fatty acids which tend to be fully saturated. By contrast, organisms which are exposed to lower temperatures have either more unsaturated acids or ones with shorter average chain lengths or a combination of both of these to give phospholipids whose tails remain fluid. The significance of the maintenance of membrane fluidity will become apparent later. The process of hardening plants against injury from frost or chilling is accompanied by changes in the degree of unsaturation of the membrane lipids.

Fatty acid analyses of individual phospholipids extracted from various sources. Bearing in mind that there is a great deal of room for manoeuvre in selecting the fatty acids to suit the prevailing environmental temperature the values for the relative abundance of fatty acids should be considered only as very general guides to the types of acid found in nature.

Thus, the predominant fatty acids have even numbers of carbon atoms, the saturated acids found most frequently are palmitic (16:0) and stearic (18:0), and the principal unsaturated acids are linoleic (18:2) and the triply unsaturated linolenic (18:3). The fatty acid composition of lecithin can depend very strongly on its origin. For example, the lecithin in the outer mitochondrial membrane is much richer in palmitic acid (16:0), and perhaps is a less fluid component than in the inner mitochondrial membrane where triply unsaturated linolenic (18:3) is the most abundant fatty acid.

Glycolipids

In several respects the glycolipids resemble phospholipids. The molecule is amphipathic, the polar group being a galactosyl derivative of a diglyceride, the non-polar part of the molecule being a pair of long, straight-chain fatty acids. Glycolipids are unusually rich in the triply unsaturated linolenic acid ($C_{18:3}$) which may make up more than 90% of the fatty acid.

They are characteristic of photosynthetic tissues since they are the major lipid component of chloroplast lamellae, largely replacing the phospholipids. Ongun *et al.*, showed that more than 80% of all of the glycolipid in leaf cells was present in the chloroplasts. The probable orientation of these molecules in the chloroplast lamellae is much like that described for phospholipids with the fatty acid tails inserted into the central region of the membrane with the polar galactosyl groups at the membrane surface protruding into the stroma. Because the fatty acid is so highly unsaturated the membranes of lamellae probably remain fluid even at sub-zero temperatures—thus any photosynthetic reaction, or molecular reorientation, which depends on membrane fluidity may have a wide temperature range in which it can occur.

A sulphur-containing glycolipid is found as a minor component of most membranes. It is known as sulpholipid and its structure and occurrence in chloroplasts was reported by Benson; the acyl groups are mainly palmitic with a preponderance of linolenic acid, thus resembling the other galactosyl lipids. Sulpholipid represents only 1% of the total lipid in most tissues and organelles but in chloroplasts it may be as much as 10–15% of the lipid.

Sterols

A number of sterols can be extracted from plant tissues and fungi as well as from isolated membrane fractions. The conventional example of a sterol of common biological origin is cholesterol; in practice plant cells contain relatively little of this sterol in comparison with animal cells. Such meagre quantitative data as is available show that sterols having 29 carbon atoms, e.g. *b* -sitosterol, are the most abundant in higher plants, while in fungi the C-28 sterol, ergosterol is often dominant. All of these molecules have an extended concertina-like configuration (known as the 'chair' or 'boat'), seem metabolically inert and are synthesized and turned over very slowly, especially in comparison with the other lipid components of membranes. Their function in membranes is not well understood but it is likely that they have an architectural role concerned with the maintenance of structure or order in the lipid domain. In this respect all of them probably function in the same was as cholesterol because Butler *et al.*, found that the structural order of bilayer membranes synthesized from lipids of ox brain tissue was stabilized equally well by cholesterol, *b* -sitosterol of plant origin and ergosterol. In the plasmalemma of the animal and plant cell there is a much higher proportion of sterols and sterol esters relative to phospholipid than in other membranes. It should be noted, however, that the membranes of intracellular organelles contain much more protein than do plasmamembranes. To some extent this protein, much of which is bonded hydrophobically to the lipid, may function in a way similar to sterol in maintaining the structural order of the membrane interior.

Proteins

In many membranes, particularly in those of chloroplasts and mitochondria proteins make up most of the weight. The proteins found are many but as a first step in classifying them *integral* and *peripheral* proteins may be distinguished. The recognition that certain proteins are embedded deeply in the lipid membrane represents a departure from the view, often advanced in earlier texts, that all of the membrane protein is located in the two peripheral bands which stain darkly with osmium and are visualised in the electron microscope. Whereas some of the protein is certainly located in this way and is probably bonded to the polar regions of the phospholipids electrostatically, it has become apparent that much protein is associated with the non-polar regions of the lipid by hydrophobic bonding. Peripherally located protein can be easily separated by washing with salt solutions or chelating agents but integral proteins are attached very strongly to the membrane and can be removed only after drastic treatment with organic solvents or detergents; even then, the isolated protein usually has some lipid attached to it.

Most of the membrane-bound enzymes, transport proteins (e.g. monovalent cation stimulated ATPase), drug and hormone receptors (in animal cells) and antigenic proteins are *integral* and are revealed when membranes are split open in freeze-fracturing. In many instances the enzymes arenonfunctional *in vitro* in the absence of lipid. It is thought that the non-polar parts of the polypeptide chains are associated with the hydrophobic tails of the fatty acids and, this being so, several types of conformation are possible. Proteins, like phospholipids, are amphipathic and their polar regions will arrange themselves so that their contacts with the hydrophobic regions of the membrane will be minimized; to ensure this, a protein could be arranged so that its polar, hydrophilic region lies among the phospholipid 'heads', or projects through them into the protein at the membrane periphery. (Fig. 1)

Long polypeptide chains with charged groups at either end may actually lie across the width of the membrane, or groups of them might lie with polar groups directed inwardly to form a hydrophilic pore across the membrane. An alternative conformation would be provided by the formation of a globular structure in which all of the polar groups would be directed towards the centre of the globule so that a hydrophobic surface would be presented to the lipid. This latter kind of conformation is probably least common.

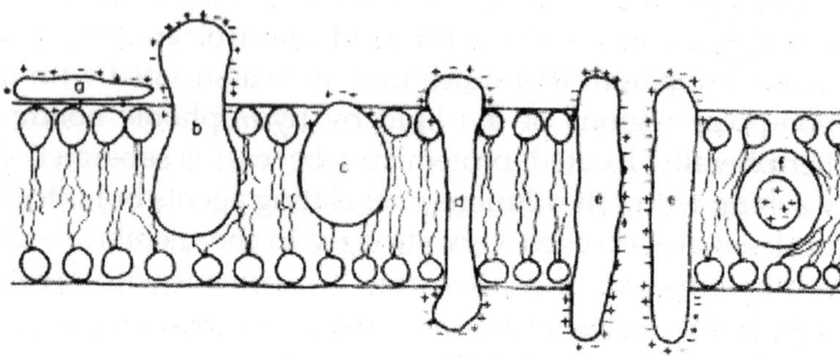

Figure 1. Possible orientations of proteins in a membrane. (a) peripherally bound protein with polar groups all over its surface. (b) and (c) non-polar regions of the protein bonded hydrophobically to lipid but with different numbers of polar groups. (d) polar groups at either end of a long molecule with a non-polar central region. (e) a pair of proteins as in (d) making up a polar pore or channel in the membrane. (f) a hydrophobic globular protein wholly in the lipid domain-polar groups, if any, directed inwardly.

The peripheral proteins can be attached to the polar groups of either phospholipids or integral proteins; examples which might be taken include cytochrome *c* which is located on the outer surface of the inner mitochondrial membranes, the chromoprotein, phytochrome, which is thought to be attached to the plasmalemma and the sulphate and other ion-binding proteins on the outer surface of bacterial membranes.

From a quantitative point of view, certain generalisations about the relative abundance of peripheral and integral proteins can be made. The greater the metabolic activity which centres upon a given membrane system the greater amount of protein integrated into it. Thus, it might be anticipated that chloroplast lamellae and inner mitochondrial membranes would be relatively rich in these proteins, whereas membranes whose role is more concerned with providing a diffusion barrier, e.g. the plasmalemma and the tonoplast, would be less so; evidence from electron microscopy shows that this is so.

Water

Water is an important, if neglected, constituent of membranes for several reasons. In a general way it determines their basic design since, in its presence, amphipathic lipids assume a bilayered configuration. There are, however, other specific associations of water molecules and membrane components which are not fully understood.

It has been estimated that water of hydration accounts for about 30% of the weight of membranes. Much of this water will almost certainly not be in a liquid state but will exist in ordered layers around the hydrophilic parts of lipids and proteins. Immobilised by hydrogen bonding these water molecules are in a liquid-crystalline condition and cannot be frozen to form ice. Water layers bound at the surface of the membrane have been estimated to have viscosity of 39 times that of pure water and to have a thickness of at least 2.2 nm. They must contribute to the mechanical stability of membranes and add significantly to their barrier properties to diffusing solutes. Expressing an extreme view, Ling has proposed that it is not lipid but these polarized multilayers of water which provide the cell with its selective surface barrier.

Hydrophobic bonding between the non-polar regions of lipids and integral proteins is favoured thermodynamically by the interactions of their polar regions with water.

Much experimental evidence points to the fact that water molecules are not restricted to membrane surfaces but cross the hydrophobic regions in numerous water-filled pores. These pores are thought to conduct water and small solutes (diameter <0.4 nm) to which membranes are highly permeable. Some water lining these pores is fully 'organised' and should, therefore, be regarded as a structural feature but there is indirect evidence to suggest that some of it must be free water *in transit*.

MEMBRANE STRUCTURE

It is convenient to discuss membrane structure under two headings; the organisation of the *membrane matrix* which is largely a matter of the relationships of the lipid components, and the *substructure* of the protein in the membrane.

Membrane Matrix

Phospholipids

The structure of phospholipid molecules considered earlier provides the key to understanding why it is that the membranes as seen in transverse sections have a characteristic trilaminar appearance. The hydrophilic head regions make hydrogen bonds with water and may become cross-linked to other heads and to proteins through ionic bridges, e.g. by calcium ions; thus they are organised into a lattice-like structure. By contrast the long acyl chains of the two fatty acids attached to each phospholipid are strongly hydrophobic, loosely organised and, above their melting point, are relatively fluid.

If phospholipid is dispersed in water the 'tails' will take on a conformation which will minimise their contacts with water. The 'heads' will, of course, react favourably with water. If the available water surface is large relative to the amount of lipid, the molecules will arrange themselves as a film-like monolayer with the heads at the water surface and the 'tails' protruding from it at right angles. (Fig. 2)

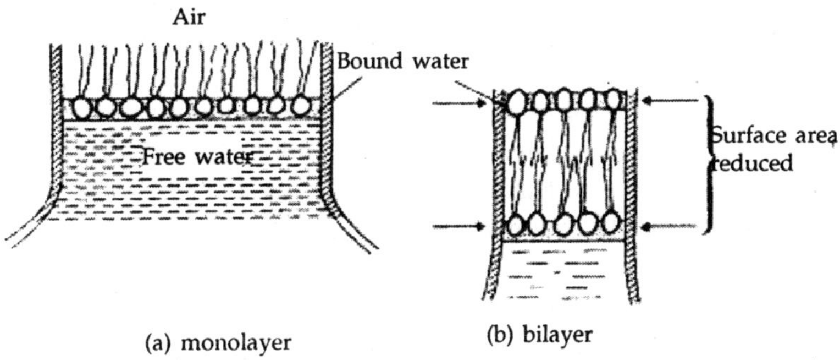

Figure 2 . An illustration of how a monolayer of dispersed phospholipid (a) in water, forms into a bilayer, (b) on contraction of the water surface area.

If more phospholipid molecules are added to this system so that there are more than can be fitted into a tightly packed monolayer over the water surface, a second type of arrangement occurs quite spontaneously. The phospholipids form two ranks with the heads facing outwards in both and the tails directed inwards to form a non-polar hydrophobic layer sandwiched between them. This bilayer arrangement, which is common to all biological membranes, can also be formed from mixtures of phospholipids under laboratory conditions. The synthetic membranes thus produced have helped in arriving at an understanding of many of the structure/function relationships of natural membranes.

The selected analyses in Table 1 show that a given membrane may contain several types of phospholipid as well as appreciable quantities of sterol. It is probable that there is a great deal more organisation of phospholipids in natural membranes than can be demonstrated positively at present. Lipids of one kind may be associated into clumps so that the membrane surface may be very heterogeneous with lipids of differing physical properties arranged in a mosaic.

A mosaic of charged and uncharged areas might occur because some phospholipids carry a net electrostatic charge at normal pH values, e.g. phosphatidyl glycerol, while others, like phosphatidyl choline (lecithin) are neutral. This is of significance because it has been shown that, in synthetic bilayers, the surface charge on the phospholipid heads can partly determine both the ion-selectivity and the cation permeability of the membrane, and it may also be relevant in determining regions of the surface of the plasmalemma where endo-cytosis may occur.

Local variations in the packing of sterols may render some parts of the membrane less fluid than others and thus determine areas where diffusion may be severely restricted.

More recently researchers have begun to investigate the possibility that the inner and outer halves of the bilayer may differ in their phospholipid composition. Should this prove to be the case, then it is possible that the barrier properties of the membrane to solute diffusion may be different when the membrane is approached from different sides.

In some special circumstances phospholipid molecules may become arranged into globular micelles in which the polar groups are directed towards the periphery of the sphere, the surface being hydrophilic. This state of affairs can be induced by dehydration in synthetic membrane systems, and in nature by viruses which create membrane instability, e.g. sendai virus, and by certain phospholipids (e.g. lysolecithin) with wedge-shaped head regions which tend to induce curvature of layers of closely packed phospholipids when they are introduced into a bilayer. It has been suggested that rapid local transitions from the predominant bilayer to the micellar state are important in membrane fusion and in pinocytosis. If these transitions do occur then it is possible that they may cause transient gaps or pores to be created in the membrane; much physiological evidence points to the conclusion that membranes do have very fine pores in them.

Sterols

The insertion of sterol molecules into the membrane increases the structural order of the hydrophobic region. These molecules lie with their long axes parallel to the hydrocarbon chains of the fatty acids with their more rigid ring structures directed towards the outside and their open chain ends towards the centre. The mobility of the hydrocarbon chains nearest to the outside of the membrane is, therefore, restricted by these stiffening structures but they remain pliant at their ends so that the central region is fluid. The rigidity conferred on the membrane by the inclusion of sterols slows down the diffusion of materials through the outer part of the lipid domain in synthetic bilayers.

Membrane Sub-Structure

When a cell or a piece of tissue is frozen and then fractured with a suitable blade, the fracture plane will follow lines of weakness in the structure. Since the central region of the membrane matrix contains no ice it is a potential line of weakness, along which the membrane tends to fracture. Where the fracture line passes tangentially across a cell or organelle, sheets of membrane material become apparent; since membranes tend to be cleaved down the middle it is obvious that the surface exposed is not the true membrane surface but is the membrane interior. Some of the particles are arranged in files while others are randomly distributed. These particles, which are usually 6–9 nm in diameter, are embedded in the membrane and are not resting on its true surface. This was demonstrated by Pinto da Silva & Branton who etched away the ice from the fracture plane by leaving the specimens under a high vacuum for some minutes after fracturing them. Using this method, the true surface of membranes which lay obliquely to the fracture plane was eventually revealed as the ice from the surrounding cytoplasm sublimed. The true surface had a much smoother appearance, the undulations of which gave the impression of a blanket lying over the embedded particles. The particles in the membrane can be removed by treatment with proteolytic enzymes and can be re-created in synthetic membranes which have been made in the presence of a hydrophobic protein. Vail *et al.,* eported that a synthetic bilayer membrane into which a hydrophobic protein had been incorporated had intercalated particles of 8.5 to 9 nm diameter occupying 12% of the internal membrane surface whereas in bilayers lacking the protein there were none.

The frequency of particles varies according to membrane type; chloroplast and mitochondrial membranes have the highest frequency as one might anticipate from the many metabolic events which are centred upon them. One of the two fracture faces is usually more densely populated than the other. Bearing in mind the amphipathic nature of membrane proteins it is possible that the more densely populated face may reflect the principal orientation of the polar groups of the integral proteins, i.e. if most of the polar groups were directed towards the outside of the cell or compartment, the outer fracture face would be the most densely populated. It should be noted that in synthetic lecithin membrane and in nerve myelin, which both have a very low permeability to ions, there are no particles.

Membrane Fluidity

An essential feature of the model in Fig. 3 is that the various components of the membrane can move relative to one another. Rather than being fixed points, many of the proteins are best thought of as drifting around in the lipid, the viscosity of which will determine the rate at which they move. Evidence that membrane particles can move in the plane of the membrane comes from freeze fracture studies of the plasmalemma of *Mycoplasma mycoides*, in which it was seen that particles became aggregated as the membrane was cooled down but dispersed again when it was warmed up. A second line of the evidence comes from studies where two animal cells were induced to fuse with one another. One of the cells has a specific antigen in its plasma membrane which the other cell lacked.

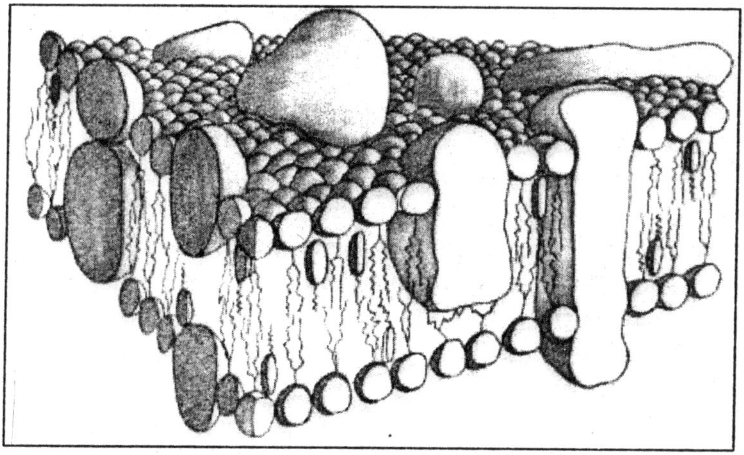

Figure 3. A representation of the fluid mosaic model of membrane structure showing peripheral and integral proteins in a 'sea' of lipid molecules.

The distribution of the antigen could be visualised by the binding of a fluorescent antibody. The fusion of the cells to form a heterocaryon and the subsequent redistribution of the fluorescent antibody were observed using a fluorescence microscope. At first the fluorescent marker was restricted to one half of the heterocaryon, but quite quickly, and in an orderly progression the fluorescence spread right around the coat indicating that the antigenic protein, known to be integrated into the membrane matrix, had diffused in the plane of the membrane. For it to have done this it must have drifted through the lipid. If the heterocaryon was cooled below the transition temperature of its lipids, no mixing of the antigen occurred.

If the membrane is cooled to the extent that the lipids freeze then such diffusion will become very restricted; the biological significance of this is reflected in the fact that the rates of most physiological processes examined over a range of temperatures are found to have a sharp transition at a temperature which is close to the transition of membrane lipids from a liquid crystalline condition to the gelled condition.

Most workers agree that there must be some proteins whose position in the membrane relative to others must be maintained, e.g. components of electrontransport chains. Such proteins may require anchoring points either in parts of the membrane which are less fluid than others, or by association with some extra-membrane protein. It is possible that aggregations of sterol molecules might serve to create more viscous patches, and there is evidence, again from .synthetic lipid bilayers, that cholesterol can be concentrated in association with certain phospholipids.

TRANSPORT OF SUBSTANCES ACROSS MEMBRANES

In a healthy cell there is a continuous interchange of water, ions, uncharged solutes, metabolites and dissolved gases across the plasmalemma. As one might expect, not all of these substances move through the membrane in the same manner. Firstly, some substances diffuse into a cell down a gradient of potentia energy; such movement is spontaneous and is the thermodynamic equivalent 0[1] heat passing from a warmer to a cooler body. There are, however, many substances which are accumulated by cells against a gradient of potential hence their movement into the cell is 'uphill' and is equivalent to the flow of heat from a cooler to a warmer body. 'Uphill' transport requires work to be performed and thus consumes energy. 'Downhill' transport is frequently described as passive while 'uphill' transport is described as active and directly involves the participation of cellular metabolism.

Correlation of Structure and Function

Having a picture in mind of the way membranes are assembled and of the driving forces which operate across them it is now possible to consider how their design, particularly that of the plasmalemma, manages to reconcile two conflicting sets of priorities. This conflict may be summarised as follows: the cell usually contains solutes at a far higher concentration than in the external medium; without any barrier these solutes would disperse by diffusion. Many solutes are accumulated by mechanisms which consume metabolic energy and it is sensible for nature to minimise their subsequent leakage and thereby economise in the use of 'fuel'. Insulation of this type might be most effective if the cell were surrounded by a barrier impenetrable to virtually everything except water and dissolved gases.

Such a barrier would, however, isolate the cell from the outside world and render it insensitive to stimulae and incapable of growth, it might eliminate its interactions with other cells in adjacent or remote tissues and prevent the excretion of harmful excesses of substances generated during metabolism. The cell membrane must provide a balance, therefore, between minimising losses and permitting a selective interchange of substances between the cell interior and the surroundings. The balance is achieved, in many cases, by providing channels or other more elaborate carrier mechanisms whereby materials crossing the membrane may essentially by-pass the hydrophobic region.

Membrane Pores and Channels

Lipid or oil is a very effective electrical insulator (it is used in high voltage underground electric cables for this purpose) and it is not surprising to find that the electrical resistance of synthetic bilayers made from pure phospholipid is very high, being 10^7 to 10^9 ohms cm^2 . In nature, the current conducted across membranes is carried by ions and we can see that an unmodified lipid bilayer with such a high resistance is a poor material across which to conduct this essential current of electrolyte. It is not surprising to find, therefore that the electrical resistance of natural membranes is very much less than that of bilayers, usually lying in a wide range of 10^2 to 10^5 ohms cm^2

Certain antibiotic molecules, such as nystatin, appear to condense cholesterol molecules in both synthetic and natural membranes to form pores with a radius of *ca.* 0.4 nm. The presence of such pores greatly increases the electrical conductance and hydraulic conductivity of the membrane. In passing we might note that many antibiotics have their effect by enormously increasing the passive permeability of cell membranes causing non-resistant cells to lose their contents or to lyse. Interest in these substances stems from the experimental evidence that cell membranes also possess pores of similar size and that the antibiotic merely induces an extreme expresssion of the normal condition.

Although the word 'pore' is often used to describe channels through which solutes and water can move, we should resist the temptation to conclude that all 'pores' are definable structural entities like the ones induced by nystatin. In many instances a 'pore' may be more like a transient imperfection in membrane structure. This latter type may have a certain statistical probability but have no fixed position.

Proteins which traverse the lipid layer may give rise to hydrophilic channels or pores. Indirect evidence supporting this idea comes from a study in which red blood cell membranes were exposed by deeply etching frozen cells under vacuum. Shrinkage of the membrane surface was observed in areas overlying groups of membrane particles possibly due to sublimation of ice from within the embedded particles; for this to have occurred this water would be a free liquid in the thawed condition.

The first circumstantial evidence for the existence of pores come from studies by Collander and Barlund on the permeation of the giant internodal cells of the alga, *Chara ceratophylla*, by a number of uncharged solutes and water. They found that in almost every case the rate of movement of a substance into a cell depended on its molecular weight and dimensions and on its solubility in oil relative to water; substances with high oil solubility permeated most rapidly. This strongly suggested that movement across cell membranes involved the movement of the solute out of the water, its solution in lipid and subsequent diffusion through it, and its re-entry into the aqueous phase at the inside face of the membrane.

The upper size limit for molecules which behaved anomalously was a radius of 0.4 nm and it was suggested that the membrane was constructed as a very fine sieve containing pores of 0.4 nm radius through which water (0.25 nm radius) and certain solutes could move. Since this early work a great deal has been done and the equivalent pore radius in many plasma membranes has been confirmed as 0.4 nm. It must be said, however, that some authorities are reluctant to accept that pores can provide channels for the bulk movement of water and solutes; the arguments for and against pores have been clearly discussed in Oschman *et al.*

Membranes are known to control very precisely the relative rates at which various ions will diffuse across them; potassium ions will diffuse ten to one-hundred times faster than sodium. Ions in solution are hydrated by binding one or more water molecules; it requires a great deal of energy to dehydrate an ion and for this reason we should think of ions in all natural circumstances as being hydrated. Sodium binds 5 water molecules, whereas potassium binds 3, the former is, therefore, the more bulky ion whose diffusion into the narrow water filled pores would be slower than for the smaller hydrated potassium ion. On the other hand the anion chloride, which has only one water molecule in its hydration shell, diffuses much more slowly than either K^+ or Na^+ . This is probably due to the fact that 'pores' carry a predominantly negative charge so that cations would be attracted to them, while anions would be repelled—a small number of positively charged pores would handle the flow of anions. Polyvalent cations and anions have much more water in their hydration shells, e.g. Ca^{2+} has 10 and SO_4^{2-} has 8 and it seems likely that these would be totally excluded from the pores.

Ion Pumps and Membrane Substructure

It is most unlikely that the intercalated membrane particles described earlier are simple ion-carriers—they are much too large—but they may be ion pumps, or groups of pumps. The best evidence for this assertion comes from work on the ATPase which pumps Ca^{2+} across the membranes of sarcoplasmic reticulum. Racker was able to isolate this enzyme and put it back into a synthetic bilayer membrane made from soybean phospholipids, and thus reconstitute a membrane which could actively transport Ca^{2+} when ATP and Mg^{2+} were present in the medium. Working with a slightly different system

Packer *et al.*, showed that freeze etched fractures of both the original tissue membrane and reconstituted membrane containing the purified ATPase were densely studded by numerous particles 8.5 nm in diameter. There can be little doubt that the particles were the calcium-pumping ATPase. Similar, less complete, evidence is also available for the Na^+ /K^+ -dependent ATPase of the red blood cell. There is no insuperable difficulty in repeating such observations with plant tissues but, at the time of writing there is no report that this has been done.

One should not conclude that all of the membrane substructure is concerned with solute transport, especially in mitochondria and chloroplasts where other biochemical activities are centred on membranes.

Endocytosis and Vesicular Transport

The presence of abundant vesicles in the vicinity of the plasmalemma of plant cells has been pointed out earlier; however, it is not yet clear what contribution endocytosis makes to the total solute transport across this, or any other membrane. There is kinetic evidence from several sources which is consistent with the notion that a measurable fraction of various ions in the cytoplasm of plant cells is sequestered into a compartment separate from the bulk of the cytoplasm. MacRobbie found that labelled Cl^- transported into the vacuole of *Nitella translucens* could be easily resolved into a fast and a slow component. The fast component was envisaged as being delivered to the vacuole formed in vesicles which had been found at the plasmalemma. The chloride ions in these vesicles would not have mixed with the unlabelled chloride in the bulk cytoplasm, whereas labelled Cl^- delivered directly to the cytoplasm *via a* pumping mechanism would have to mix with a much larger pool of unlabelled chloride ions. At the beginning of the experiment, therefore, the labelled Cl^- in the vacuole increased more rapidly than expected.

If the vesicles contained nothing more than a small volume of the outside solution, their net contribution to the solute content of the cell would be very small and they could not exercise the ion-selectivity which characterises cell membranes. Baker and Hall suggest that endocytosis becomes a much more plausible transport process if it is assumed that ions become selectively bound to the membrane surface and thus become concentrated from the dilute external medium. This assumption is entirely reasonable because both proteins and phospholipid heads can provide ion-selective binding sites. These authors also relate membrane-bound ATPase activity with the subsequent invagination of the plasmalemma to form a vesicle but this implies that vesicle formation is energy-dependent. This is against the substantial evidence from animal cells which shows that micro-vesicle formation results from Brownian movement of the membrane and does not need to be energised by ATP; ATP is required for the very large vesicles formed by re-orientation of micro-fibrils in phagocytosis, but the type of vesicle under discussion here is much smaller, probably 70–100 nm in diameter and is not dependent on microfibrils. There is visual evidence that these micro-vesicles can discharge their contents directly into lysosomal vacuoles.

The important advance of thinking required to deal with endocytosis will provide a severe strain on biophysical theorists accustomed to thinking of cells as homogeneous phases separated by diffusion barriers, partly because materials can reach the interior of the cell without crossing a membrane at all. Although endocytosis is probably a common phenomenon it is likely to remain an 'unknown quantity' until new experimental techniques are devised.

REFERENCES

Anderson W.P. (ed.), *Ion Transport in Plants*, Academic Press, London and N.Y., 1973.

Branton D. & Deamer C.W., *Membrane Structure*, Springer-Verlag, New York/Wien, 1972.

Eisenberg M. & McLaughlin S., "Lipid bilayers as models of biological membranes", *Bioscience*, 1976.

Lucy J.A., *Lipids and membranes*, FEBS Lett. 40, S105-S111, 1974.

Raven J.A. & Smith F.A., "Significance of hydrogen in transport in plant cells", *Canadian Journal of Botany*, 1974.

12

Protein Synthesis in Plants

Living organisms make several thousand different proteins with a variety of functions, present evidence suggests that they are all made, by the same basic mechanism, on polysomes. These structures consist of a template RNA molecule, messenger-RNA (mRNA), with several associated ribosomes, and protein synthesis is a complex, energy requiring, multi-enzymic process. A template mechanism of synthesis is dictated by the fact that the protein product has a specificity which depends upon the precise sequence of the constituent amino acid residues in the polypeptide chain. This automatically rules out the normal enzymic method by which most other cell constituents are made, since each enzyme of a set specific for the synthesis of one protein would, in turn, require a further set for its synthesis, *ad infinitum*. However, since some small polypeptides are synthesized by a non-template mechanism, it is possible that some proteins with highly repetitive sequences of amino acid residues, may be synthesized similarly.

Because all organisms probably synthesize proteins by a similar mechanism, the tendency has been to use an organism which is particularly amenable for basic studies. This organism is the prokaryote, *Escherichia coli*, and higher plant material has not been much used because of associated technical difficulties. However, recognition of the importance of the deposition of proteins in plants as our major source of protein food, is leading to an increased spate of work in the desire to understand, and possibly manipulate, this important process in the major food plants.

POLYSOMES

Structure of Ribosomes

Ribosomes are classified as eukaryotic or prokaryotic in type, on the basis of: (a) sensitivity to various antibiotics; (b) functional interchangeability of soluble factors and ribosomes from different sources; and (c) structure and sedimentation characteristics, e.g. 70s (prokaryotic) or 80s (eukaryotic).

Bacterial ribosomes, plant chloroplast and mitochondrial ribosomes are classified as prokaryotic, even though they may not always exhibit all of the above characteristics, e.g. plant mitochondrial ribosomes have sedimentation values of about 80s. Plant and animal cytoplasmic ribosomes are of the eukaryotic type and have sedimentation values of about 80s. This class of ribosomes are heterogeneous in size however, with most of the differences accounted for by differences in the large subunit, the small subunit having been conserved during evolution.

Measurements of the size of plant cytoplasmic ribosomes in the electron microscope vary, but they are approximately 25 × 20 nm. Miller *et al.*, have described them as being acorn-shaped, and several workers have described a cleft in the small subunit. In the rat liver ribosome model of Nonomura *et al.*, there is a tunnel between the two subunits, directly under the cleft in the small subunit, which is thought to accommodate the mRNA.

Ribosomes require Mg^{2+} for structural integrity and dissociate at low Mg^{2+} concentrations into a large and a small subunit with sedimentation values of about 60s and

40s respectively. There is good evidence that mRNA attaches to the small subunit before the addition of the large subunit completes the ribosome structure, and the different roles of the two subunits are a feature of all the models of ribosome structure and function.

The 80s ribosome of plants corresponds to a molecular weight of 3.9×10^6 daltons, with molecular weights for the large subunit of 2.4×10^6 daltons and the small subunits, 1.5×10^6 daltons; the small subunit is of the same molecular weight as its mammalian counterpart, the large subunit is smaller. The large subunit of plant ribosomes contains one molecule of 25s RNA, hydrogenbonded to 1 molecule of 5.8s RNA, and 1 molecule of 5s RNA; the small subunit contains 1 molecule of 18s RNA. The molecular weight of the 25s RNA is 1.3×10^6 daltons, and that of the 18s is 0.7×10^6 daltons. The large subunit contains 46% protein, and the small 54%.

The large subunit of pea seedling ribosomes contains 44–45 proteins and the small subunit 32–40 proteins, of which most are basic and of molecular weights between 20×10^3 and 30×10^3 daltons. Some proteins may be represented by more than one copy per ribosome. Proteins extracted from cytoplasmic and chloroplast ribosomes of the same species, show little similarity in two-dimensional electrophoresis or by immunological comparison and are significantly less similar than are the cytoplasmic ribosomes of different species, e.g. beans and wheat.

Before a complete understanding of the mechanism of protein synthesis is elucidated, it will be necessary to know the spatial relationships and three-dimensional structures of the proteins and the RNAs of the ribosome, as well as those of the associated molecules which also play a part in protein synthesis. Information on the structure of the proteins and the RNA molecules of the *E. coli* ribosome, together with reconstitution experiments, is now well advanced.

Free and Membrane-Bound Polysomes

Ribosomes associate with mRNA to form polysomes, as can be seen in the electron micrograph; the size of the polysome varies according to the length of the mRNA and the number of attached ribosomes. Polysomes are found either free in the cytoplasm or attached to the surface of membranes of the endoplasmic reticulum (ER) and the nucleus. In animal embryonic cells, most of the polysomes are free and engaged in protein synthesis for internal use, whereas in those differentiated animal cells from which large amounts of protein are exported, most of the polysomes are found attached to the ER. Thus, the generalisation arose that membrane-bound ribosomes synthesize protein for export and free ribosomes for intracellular use. It soon became clear however, that membrane-bound ribosomes occur in some tissues which do not export protein, and that in cells where all the protein synthesized is for internal use, different classes of protein are synthesized on the two types of ribosomes. Less information is available for plants.

In the developing broadbean seed the highly vacuolate, relatively membrane-free cells of the cotyledon are transformed just prior to the onset of storage protein synthesis, into cells whose cytoplasm is vesicular and in which ER with attached ribosomes is very prominent. This new protein synthesis machinery is assembled at a precise time in the course of seed development for the production, in large amounts, of the storage proteins of the seed. Later, during dehydration of the seed, ribosomes become detached and the population of free ribosomes thereby increased. A similar series of events has been recorded for the developing seeds in many other plants.

The way in which ribosomes attach to membranes is not clear; the ribosomes themselves, the membranes and the protein being synthesized, have all been suggested as being involved in the binding. There is evidence from animals that the large subunit is in close contact with the membranes, and Baglioni *et al.* have postulated that the large subunit binds directly to the membrane, probably at a specific binding site, and that a protein on the membranes is responsible for the attachment. Alternatively, it has been suggested that one of the proteins of the large subunit is responsible, whereas other observations suggest that the binding may be dependent on the nascent polypeptide chain. It now seems likelly that, *in vivo*, membrane-bound ribosomes synthesize a different class of protein to free ribosomes, whereas *in vitro*, proteins of both classes are synthesized by both types of ribosomes, suggesting the control is not a function of the ribosome itself.

Isolation and Purification

Ribosomes and polysomes are normally isolated from plants and purified by sedimentation through sucrose cushions as originally described by Wettstein *et al.* In a typical procedure, the material is homogenised in 0.25 M sucrose containing 200 mM -Tris-HCl, pH 8.5 at 2°C, 500 mM -KCl and 15 mM -$MgC1_2$, with a Willems Polytron for 3 seconds at a speed setting of 8. The homo-genate is filtered through Miracloth and Triton X-100 is added to a final concentration of 2% (v/v). The filtrate is centrifuged at $10^4 \times g$ for 10 min at 2°C and the ribosomes and polysomes are recovered from the supernatant by layering over a 3 ml cushion of 0.1 M sucrose in 50 mM -Tris-HCI, pH 8.5, 50 mM -KCl and 10 mM -$MgCl_2$, followed by centrifugation for at least 44 x 10^7 g-min, as described by Leaver & Dyer (Fig. 1). The inclusion of the detergent Triton X-100 is necessary since plant materials contain a proportion of polysomes which are membrane-bound and the addition of Triton X-100 solubilises the endoplasmic reticulum, so releasing them.

The method gives a preparation of both polysomes and ribosomes and the latter can be removed by centrifugation techniques or, alternatively, the whole preparation can be converted to ribosomes by exposure of the plants to nitrogen gas for at least 1 hour prior to extraction. Since different plants and even different tissues from the same plant contain different amounts of membrane-bound to free polysomes and different free polysomes to free monoribosome ratios, isolation conditions may need to be varied for optimum results with different experimental materials.

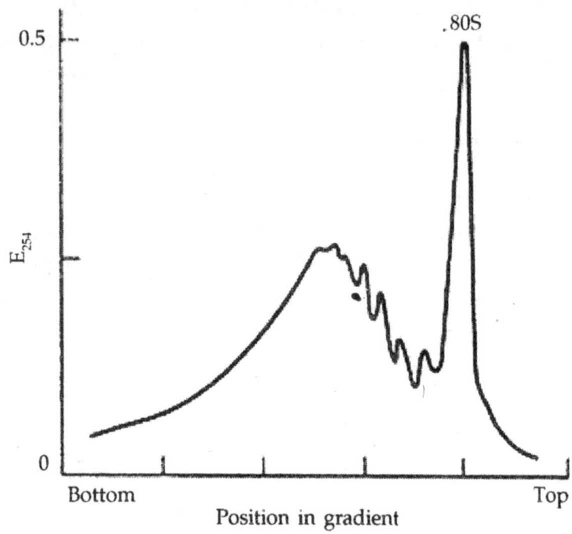

Figure 10.3 . Sucrose density profiles of ribosomes and polysomes from Pisum sativum.

The situation is further complicated by the fact that plant mitochondrial ribosomes, although not functionally of the 80s type, sediment at 80s, and can contaminate cytoplasmic preparations to varying extents, depending on the experimental material. Damage to ribosomes, both structural and by the removal of associated proteins, can occur during isolation and preparation. The extent of this damage can be assessed, to some extent, by extracting the

RNA and fractionating it by polyacrylamide gel electrophoresis.

BIOCHEMICAL MECHANISM OF PROTEIN SYNTHESIS

Protein synthesis, i.e. the assembly of polypeptide chains from their amino acid constituents, occurs on the ribosomes. The information specifying the amino acid sequence of the polypeptide chain is contained in mRNA molecules, which in the case of eukaryotes are complexed with specific proteins *in vivo.* Messenger-like RNA-protein particles called 'informosomes' have been isolated from the cytoplasm and from polysomes of germinating wheat embryos. So far, however, the RNA of these particles has not been definitely established as mRNA. When an mRNA molecule associates with ribosomes, the information it contains is translated; the number of ribosomes involved depends largely upon the size of the polypeptide chain being synthesized, since generally, ribosomes are evenly spaced along mRNA molecules.

The translation process can be arbitrarily split into initiation, elongation, and termination and release. However, prior to the initiation of the process on the ribosome, the twenty different protein amino acids must be activated and attached to specific tRNA molecules. Also, after release of the polypeptide chain from the ribosome, a series of post-translational changes may occur.

tRNA

The code words (codons) of the genetic code were first established for *E. coil.* This work showed that there were 61 codons shared between twenty protein amino acids, i.e. there is more than one codon for some amino acids and these can attach to more than one tRNA species. Chemically different tRNA species which can be acylated by the same amino acid are called isoacceptor tRNAs.

However, it was soon realised that not every codon has a corresponding tRNA with a specific anticodon and that some tRNAs recognise more than one codon. Crick's 'wobble hypothesis' to account for this, proposed alternative base-pairing to occur between the base in the third position of the mRNA codon and the corresponding base in the tRNA anticodon. The proposed rules are set out in Fig. 2.

5'-Anticodon base	mRNA codon
I	XXU, XXC, XXA
G	XXU, XXC
U	XXA, XXG
A	XXU
C	XXG

Figure 2. Basis of the 'wobble' hypothesis.

Thus, inosine in the 5' position in the anticodon of an aminoacyl-tRNA molecule allows this base to pair with either U, C or A in the third 3' position of a messenger codon. Similarly, with G or U in the 5' position in the anticodon, two codons may be recognised by a single aminoacyl-tRNA, whereas with C or A only a single codon is recognised. Since alternative base-pairing is only possible with the base in the third position of the codon, codons for the same amino acid which differ in either of the first two base positions, base-pair with different tRNAs. The only example of 'wobble' in the first base of a codon occurs with $tRNA_i^{Met}$, which recognises both the initiator AUG or GUG codons.

Even so, separation of tRNAs by counter current-distribution, MAK columns, BD-cellulose columns and reversed phase chromatography has shown that there are more isoacceptor tRNAs than can be accounted for by the degeneracy of the genetic code. Similarly, isoacceptor tRNAs for amino acids exist in plants; for example, soyabean seedlings have at least six different $tRNA^{Leu}$ species. Although few experiments have been carried out, it has been found in every case investigated that plants use the same codon in amino acid assignments as those of *E. coli*. However, plants may not use all of the possible codons for a particular amino acid, since some isoacceptor tRNA species may be absent, or because of the 'wobble' effect; e.g. Caskey *et al.*, have shown that the isoleucyl tRNA from guinea pig liver recognises AUU, AUC and AUA codons, whereas the corresponding isoleucyl-tRNA from *E. coli* only recognises two of these, AUU and AUC. This response to different codons in the two organisms is explained if the liver tRNA had 5' inosine in the anticodon, while *E. coli* tRNA had guanine.

The whole question of the number and the activity of tRNAs in different tissues of plants is complicated by the fact that, (a) it is not certain whether the methods for separating tRNAs are completely effective; (b) it is known that during isolation and purification, partial modification of tRNA molecules can occur, which may affect their charging ability with amino acids or their ability to transfer amino acids to proteins, or both; (c) the relative importance of the three protein synthesizing systems, cytoplasmic, chloroplast and mitochondrial, each with its associated tRNAs and synthetases may differ in different tissues. Nevertheless, different complements of isoacceptor tRNAs may occur in different tissues and changes in pattern have been demonstrated during development, and it has been suggested that isoacceptor tRNAs are involved in the control of protein synthesis. Garel *et al.*, have shown that the complement of tRNAs synthesized in the silk gland is related to the mRNAs being translated and they have suggested defining this phenomenon as 'modulated tRNA biosynthesis'.

Aminoacyl-tRNA synthetases

Aminoacyl-tRNA synthetases have been isolated and purified from a variety of plants, and although a considerable amount of work has been done on their properties, several problems remain. For example, the question of the detailed substrate specificity of different aminoacyl-tRNA synthetases is still under investigation. Since activation is a two-step process, both amino acid and tRNA specificity have to be considered. The synthetase is normally absolutely specific towards the tRNA; few mismatches have been reported in homologous systems. However, a synthetase specific for one amino acid can, in some instances, activate to a lesser extent another amino acid. Even so, synthetases are remarkably amino acid specific. A second amino acid attached to a tRNA molecule normally specific for another amino acid, will be located in the polypeptide chain according to the tRNA anticodon, i.e. where the first amino acid would have been placed. With regard to amino acids not normally found in the cell, specificity is not always so pronounced, e.g. azetidine-2-carboxylic acid is activated and transferred to $tRNA^{Pro}$ by the proline enzyme of mung bean, whereas *Polygonatum multiflorum*, which contains a high level of this amino acid in nature, contains a prolyl-tRNA synthetase which will not activate it. In *Mimosa* and *Leucaena*, where mimosine, an analogue of phenylalanine, occurs naturally and where it is non-toxic, the phenylalanyl-tRNA synthetase discriminates against mimosine. In several other species where mimosine is toxic, it is activated by the phenylalanine-tRNA synthetase but the enzyme does not attach it to $tRNA^{Phe}$.

It is not clear if there is one or more than one synthetase whenever there are a number of isoacceptor tRNAs. In soybean seedlings where there are at least three different leucyl aminoacyl-tRNA synthetases which have different specificities towards the six leucine isoacceptor tRNAs, it would appear that there is more than one cytoplasmic leucyl-tRNA synthetase. However, in some instances, a single aminoacyl-tRNA synthetase can recognise different isoacceptor tRNAs. When both cytoplasmic and chloroplast tRNA species are present, there are probably at least two sets of aminoacyl-tRNA synthetases. In light-grown *Euglena* for example, there are two aminoacyl-tRNA synthetases for each of phenylalanine and leucine, whereas dark-grown *Euglena,* which is without chloroplasts, contains only one; there are various reports of chloroplast aminoacyl-tRNA synthetases which only acylate chloroplast tRNAs and not their cytoplasmic counterparts and *vice versa.* However, in heterologous systems such as chloroplast tRNA plus cytoplasmic activating enzymes, specificity may range from none to complete, although the biological significance of these findings is not clear.

There is evidence that, as well as the tRNAs, the complement of aminoacyltRNA synthetases may change during the life cycle of plants.

Translation of mRNA

Initiation of the Polypeptide Chain

Initiation of protein synthesis on plant cytoplasmic ribosomes starts with the assembly of a mRNA•40s subunit complex which requires ATP for its formation and this step is followed by the addition of methionyl-$tRNA_i$ to form a mRNA•40s subunit•methionyl-$tRNA_i$ complex. The process requires GTP and at least two initiation factors. Initiation is completed by the addition of the 60s subunit and the 80s ribosome so formed is then able to accept the next aminoacyl-tRNA, thus starting the elongation process. The $tRNA_i^{Met}$ is a special tRNA charged with unformylated methionine, which can enter the so-called 'P' site of the ribosome complex and base-pair with the initiator codon, AUG or GUG in the mRNA. Although mRNAs are translated from the 5'→ 3' end, evidence from sequence studies of bacteriophage messages shows that the initiator triplet is not at the 5' terminus, but is set a considerable number of nucleotides in. For example, the first initiator codon of bacteriophage R17 RNA is preceded by 91 nucleotides which are not translated. In the case of polycistronic messengers, one or more initiator triplets will occur intramolecularly.

Evidence that plants have a special type of initiator tRNA, $tRNA_i^{Met}$, comes from the wheatgerm system, and from *Vicia faba*, where it has been shown that two major and one minor $tRNA^{Met}$ species are present. The minor and one of the major $tRNA^{Met}$ species function as chain initiators as shown by AUG-dependent binding, and by N-terminal analyses of either the product of tobacco mosaic viral RNA-directed or poly-AUG, poly-GU and endogenous messenger-directed incorporation. Neither of the major $tRNA^{Met}$ species from either wheat or beans is formylated, in contrast to the initiator $tRNAi^{Met}$ of prokaryotes, and it is presumed that these are the cytoplasmic $tRNA^{Met}$s for initiator and internal methionine residues. The minor, formylatable $tRNA^{Met}$, is presumed to be the initiator of chloroplast protein synthesis.

The hydrolysis of ATP precedes the formation of the first peptide bond and may be required for recycling of the methionyl-$tRNA_i$ binding factor. The requirement for ATP has also been demonstrated in the rabbit globin synthesizing system, but not in those of prokaryotes. The order in which the initiator tRNA and mRNA bind to the small subunit in mammalian and prokaryote systems is in dispute; the order suggested above for wheatgerm by

Marcus and his coworkers, therefore, whilst agreeing with some workers on those systems disagrees in particular with the suggestion of Legon *et al.*, and Schreier and Staehelin that the initiator binds prior to the mRNA in rabbit globin synthesis.

Knowledge of the two plant initiator factors involved in binding methionyl-tRNA$_i$ and mRNA is much less complete than for the similar factors in bacterial and mammalian systems; details of the latter are, therefore, relevant. In bacteria there are three factors, IF-1, IF-2 and IF-3 involved in the initiation process. Factor IF-1 has a molecular weight of 9,400 daltons and is the most basic of the three proteins. Factor IF-2 has been shown to consist of two sub-components, both active, with molecular weights of $9 \times 10^3 - 100 \times 10^3$ daltons and 80×10^3 daltons although the latter is probably derived from the former by proteolysis during purification. Factor IF-3 is a protein or group of proteins with molecular weight(s) of $21.5 \times 10^3 - 23.5 \times 10^3$ daltons. The three factors attach to the small subunit; IF-2 binds GTP and formyl-methionyl-RNA, while IF-3 binds mRNA. Factor IF-I increases the affinity of factors IF-2 and IF-3 for the 30s subunit, and is also necessary for the release of IF-2 from the ribosome.

The order of the attachment of the different factors to the small subunit and the order of the binding of mRNA and acylated initiator tRNA, is still a matter of debate. Some results suggest that IF-1 is attached before IF-2 and IF-3, whilst others indicate that it can only be attached after. Similarly, it is possible that an unstable intermediary complex consisting of formyl-methionyl-tRNA$_i$ •30s subunit•initiation factors is formed, i.e. that mRNA is bound after initiator-tRNA, or that a 30s subunit•mRNA•IF-3 complex is formed, which is then stabilised by the addition of formyl-methionyl-tRNAi and IF- I and IF-2. The next step in the process, the addition of the large subunit, does not require GTP, but GTP hydrolysis is required to release IF-2 from the 70s ribosome complex. The role of GTP hydrolysis in initiation is still not fully understood, but it is not only required for the release of IF-2. Factor IF-3 is not only required for the binding of natural mRNAs, but also functions later either to dissociate the subunits after chain termination, or to keep dissociated subunits apart until another cycle of initiation is instituted. This activity of the IF-3 factor is referred to as its DF activity.

Several proteins which form an integral part of the small subunit have also been shown to be involved in initiation. Small subunit proteins, s1, s11, s13, s19, participate in IF-2 binding and s1, s11, s12, s13, s14, in IF-3 binding; s12 is responsible for recognition of mRNA. Shine and Dalgarno have proposed a model of how the large subunit.initiation factor complex may recognise and attach mRNA. It would appear that the sequence (5') GGAGGU (3') is present in the same relative position with respect to the first translatable AUG triplet in all prokaryotic messengers so far analysed. Furthermore, the 3' proximal end of 16s RNA has the sequence GAUCACCUCCU UA (OH), so that the underlined nucleotides could potentially base-pair with the GGAGGU sequence in the mRNA.

Further complexity of the initiation process is indicated by the work of Schreier and Staehelin, who have characterised five initiation factors (IF-E_1 to IF-E_5) for mammalian protein synthesis. Their results with rabbit globin mRNA are consistent with the formation initially of a methionyl-tRNA$_1$ •IF-E_2 •GTP complex, which is bound independently of mRNA to the 40s ribosome subunit by IF-E_3. Subsequent to this, mRNA and the 60s subunit are joined by the cooperative action of IF-E_4 , ATP, IF-E_1 and IF-E_5. The binding of natural mRNA requires IF-E_4 and ATP. IF-E_5 promotes the joining of the 40s complex with the 60s subunit and IF-E_1 inhibits complex formation in the absence of mRNA binding. However, it has been suggested that the apparent need for additional factors results from deproteinisation of ribosomal subunits and mRNAs, i.e. the requirement may be for structural proteins rather than initiation factors. The little information available suggests that eukaryotic factors are not exchangeable with their prokaryotic counterparts.

Factor EIF-3, contrary to its prokaryotic counterpart, consists of a number of different sized polypeptide chains; the question as to whether there are different EIF-3s or whether selectivity is controlled by additional mRNA specific factors, is still unresolved. A similar uncertainty to that described for the prokaryotic system surrounds the question as to whether or not methionyl-tRNA$_i$ binds to the 40s subunit prior to mRNA.

The initiation process is extremely complex and still not fully understood. A variety of proteins are involved, some being ribosomal structural proteins and others proteins which recycle during the process. These proteins interact, changing the conformation of the ribosome and thereby allowing the different steps of initiation to proceed in a sequential and orderly manner. Different mRNAs probably differ in their rate of attachment to the ribosome and/or in their efficiency in other stages of the initiation process, so affecting the frequency and rate of translation. With proteins which contain prosthetic groups their absence may inhibit initiation, since it has been shown that lack of haemin in globin synthesis, results in the formation of an inhibitor of the binding of methionyl-tRNA$_i$ to 40s subunits. Furthermore, a variety of mRNA specific factors have been proposed called 'interference' or 'i-factors', which affect the specificity of the initiation factor IF-3. Thus, Strycharz *et al.* have identified a supernatant factor in Krebs II ascites cells, which is specifically required for the initiation of the translation of encephalomyocorditis viral RNA. However, Lodish, from a kinetic analysis of protein synthesis, has proposed a model which precludes the necessity for mRNA specific factors. He points out that these have often been observed in systems which are less active than the corresponding *in vivo* system, and that the three eukaryotic systems available which translate exogenous mRNAs efficiently *in vitro,* do so with a variety of mRNAs.

Elongation of the Polypeptide Chain

The elongation process starts once the 80s plant ribosome containing mRNA and methionyl-$tRNA_i$, has been assembled and takes place by the following repeated cycle of events, each cycle being separated into:

(a) codon-directed binding of aminoacyl-tRNA;

(b) peptide bond formation; and

(c) translocation.

The ribosome has two binding sites; a 'P' site for the binding of methionyl-$tRNA_i$ and an 'A' site where all other aminoacyl-tRNAs bind. After the initiation complex has been formed, the next aminoacyl-tRNA, i.e. the one carrying the anticodon to the next codon of the mRNA, binds at the 'A' site. Once the aminoacyl-tRNA has been bound in the 'A' site, a peptide bond is formed by a peptidyl-transferase enzyme, which is part of the large subunit, such that the 'A' site now carries the aminoacyl-tRNA joined to methionine by a peptide bond. The 'P' site now carries the deacylated $tRNA_i^{Met}$, which is subsequently removed leaving an open 'P' site, and the peptidyl-tRNA moves from the 'A' to the 'P' site, so moving the message relative to the ribosome. The 'A' site is now empty and is ready for the next aminoacyl-tRNA to enter according to the next codon in the message. This cycle of reactions requires K^+, Mg^{2+}, GTP and various elongation factors, proteins, which are found in the soluble fraction of cell homogenates.

Our knowledge of polypeptide chain elongation is most complete with *E. coli,* where three elongation factors are known, EF-Tu, EF-Ts and EF-G. Factor EF-Tu binds stepwise with GTP and aminoacyl-tRNA and this ternary complex transfers the aminoacyl-tRNA to the 'A' site of the ribosome, releasing an EF-Tu•GDP complex and inorganic phosphate. Factor EF-Ts displaces GDP to form the complex EF-Ts•EF-Tu, from which GTP displaces EF-Ts to regenerate EF-Tu•GTP. The third elongation factor, EF-G, and free GTP, are responsible for the translocation of the peptidyl tRNA from the A' to the 'P' site with the prior removal of the deacylated tRNA from the 'P' site; GTP is hydrolysed during the process. The elongation factors, like the initiation factors, recycle during protein synthesis. The protein chain grows from the N-terminal end, and probably starts to fold into its three-dimensional conformation whilst the process of elongation is proceeding.

Fewer details of the elongation factors are available in eukaryotic systems. However, two elongation factors, designated as the binding enzyme EF-1 and the translocase EF-2 have been purified from ungerminated wheat embryos. The mechanism of binding of aminoacyl-tRNA to wheat ribosomes is on the whole similar to that in *E. coli*, the main difference being the absence in the eukaryotic system of a factor with the properties of EF-Ts, i.e. EF-1 = EF-Tu and EF-2 = EF-G.

Elongation factor EF-I, which requires the presence of Mg^{2+} at low concentration, K^+ and GTP, is responsible for the binding of aminoacyl-tRNA to wheat ribosomes. It contains three active forms differing in molecular weight. The larger EF-1 forms dissociate to a single species of molecular weight 6×10^4 daltons during gel electrophoresis in the presence of SDS. The three forms observed may represent a monomer, trimer and tetramer of a single protein unit, of which the trimer seems to be the most stable form. Some experimental evidence supports the view that the light species is the active form of the EF-1 enzyme.

For example, in calf brain and wheat embryo, it has been shown that the interaction of EF-1 with GTP and aminoacyl-tRNA yields a ternary complex which contains the light form of the enzyme, i.e. molecular weight 5×10^4 to 6×10^4 daltons. It is suggested that the amino group of the esterified amino acid plays an important role in aminoacyl-tRNA binding and this differs from the bacterial system where EF-Tu can react with the deaminated product of phenylalanyl-tRNA, phenyl-lactyl-tRNA.

The second elongation factor, EF-2, is involved in translocation of the peptidyl-tRNA from the 'A' site to the 'P' site during elongation. It is a protein of molecular weight 7×10^4 daltons and requires thiol compounds for activity. Ribosomes and elongation factors from the cytoplasm of eukaryotic organisms are not functionally interchangeable with their *E. coli* counterparts. Furthermore, Perani *et al.*, have shown absolute ribosome specificity, either for 70s or 80s ribosomes, of two sets of elongation factors, T and G (EF-1 and EF-2), isolated from yeast, and for factor EF-2 isolated from the alga, *Prototheca zopfii.* It is generally accepted that elongation factors within each ribosomal type (70s or 80s) are exchangeable between eukaryotes.

Termination and Release of the Polypeptide Chain

No reliable information is available on chain termination in plants, most of the work having been done with *E. coli.* However, as there are some differences between the prokaryote and mammalian eukaryote systems, these will be mentioned briefly. In *E. coli* the elongation cycle is repeated until certain termination codons on the mRNA come into the 'A' site; these codons are UAG, UAA and UGA. At least three release factors, proteins, which recognise the terminator codons and bring about the release of the completed polypeptide chain from the tRNA, have been identified. Originally designated R-1, R-2 and S these are now called RF-1, RF-2 and RF-3. RF-1 recognises UGG or UAG, RF-2 recognises UAA or UGA, and RF-3 affects the rate of release of the polypeptide chain. In mammalian systems, only a single release factor which responds to all three codons has been identified, and furthermore the prokaryotic termination requirement for GTP appears to be absent.

In addition, several other termination factors have been proposed from different organisms. Although little is known about chain termination in plants, the presumed universality of the code suggests that similar terminator triplets and proteins are involved in plants also.

Originally, it was thought that on chain termination ribosomes dissociated into subunits and that the free ribosomes found in the cell were inactive in further protein synthesis. Noll *et al.* have presented evidence that at least some of the free ribosomes found in the cell interact with a factor and partially dissociate, so that the small subunit with bound IF-1, IF-2 and IF-3, can initiate protein synthesis.

PROTEIN SYNTHESIS *IN VIVO*

The rate of protein synthesis *in vivo* can be measured by determining the amount of protein in a plant, organ or tissue at two different times; this will give the net rate since proteins 'turnover' and the amount of protein found would represent the balance between synthesis and degradation.

Various equations describing the relationship between the synthesis and degradation of protein have been proposed but there are many technical and interpretive difficulties in this type of experiment. The turnover rate for an individual protein is not the same as the value obtained for the total protein and although the latter may be of interest agriculturally, most biochemical and physiological studies require information on individual proteins. This information has only been obtained so far with a few plant proteins, although most are thought to turnover. Possible exceptions are storage proteins which are sequestered from the metabolically active cytoplasm and stored in membrane-bound protein bodies until subsequently required at a more active phase of the life-cycle.

Organ-specific proteins have been demonstrated serologically and often one or a few proteins may predominate in an organ, e.g. ribulose bisphosphate carboxylase, which often accounts for 50% of the protein of leaves. The changing complement of proteins during development is due to changes in the type or rate of protein synthesis and/or degradation. For example, Beevers and Poulson followed changes in the protein content of pea cotyledons during the period 9–33 days after flowering. Initially, protein content increased gradually, then rapidly between 21–27 days to decline once more as the seed dehydrated and matured. Incorporation experiments with ^{14}C-leucine indicated that albumin proteins were synthesized early in cotyledon development whilst globulin synthesis predominated with increasing maturity.

During the period of rapid protein synthesis, greater amounts and a higher percentage of polysomes to monosomes were extracted than at other stages in the development of the seed. However, unlike in bacteria, where induced enzyme synthesis is widespread and the product is not 'turned over' but diluted out by a rapid rate of cell division, plants have a much slower rate of cell division and protein degradation appears to be a common method for regulating the amount, and hence the activity of proteins.

The relative amounts of different seed proteins synthesized may vary considerably in different varieties and breeding lines of a crop; for example, in maize where the major storage proteins of the seeds are prolamins and glutelins normal maize contains a relatively high proportion of prolamin, whereas in the mutant Opaque 2, the proportion of prolamin is greatly reduced. This is of particular significance since prolamin is very deficient in the essential amino acid lysine, so that seed meals obtained from Opaque 2 have a higher content of lysine and are consequently more nutritious than those from normal maize.

Protein synthesis is a concerted process involving the biochemical machinery of the cell as a whole, since it consists of a series of co-ordinated reactions, the constituents of which may be synthesized in the nucleus (mRNA, tRNA), the mitochondria (ATP), the cytosol (amino acids and aminoacyl tRNAs), the ribosomes (enzymes). Although changes in the rate of protein synthesis occur, very little is known about the mechanisms involved. Theoretically, control could be exerted at any one of the many constituent reactions, this being the pacemaker reaction for protein synthesis in that particular cell/tissue. However, the control mechanisms must allow the overall process to adjust in response to genetic, physiological and environmental stimuli, so that the rate and/or the type of proteins being synthesized can change.

REFERENCES

Allende J.E., "Protein biosynthesis in plant systems", *Techniques in Protein Biosynthesis*, vol. 2 (eds. P.N. Campbell & J.R. Sargeant), pp. 55-100. Academic Press, London and New York, 1970.

Anderson W.F., Bosch L., Gros F., Grunberg-Manago M., Ochoa S., Rich A. & Staehelin Th., "Initiation of protein synthesis in prokaryotic and eukaryotic systems", *FEBS Letters*, 1974.

Boulter D., "Protein synthesis in plants", *Annual Review of Plant Physiology*, 1970.

Haselkorn R. & Rothman-Denes L.B., "Protein synthesis", Annual Review of Biochemistry, 1973.

Mans R.J., "Protein synthesis in higher plants", *Annual Review of Plant Physiology*, 1967.

Zalik S. & Jones B.L., "Protein biosynthesis", *Annual Review of Plant Physiology*, 1973.

13

Functions of Plant Hormones

The genesis of hormone research in plants occurred with the discoveries of Charles Darwin and his son Francis, who published, almost a century ago, their book entitled *The Power of Movement in Plants'*. Their experiments dealing with phototropism revealed that an 'influence' produced in the tip of a canary grass coleoptile was translocated to the basal portion where it caused the organ to grow towards the light. Almost fifty years later Went isolated a substance with just these properties which was produced in the tip of coleoptiles and transported to the basal portion of the organ. This substance was named auxin. Ten years after Went's discovery, the structure of this material was identified as indoleacetic acid (IAA).

The nature of auxin effects on plant growth and development was studied in many laboratories in the late 1930's and 1940's. However, it was not until World War II that the selective action of indoleacetic acid served as a basis for developing new chemicals for the control of growth in specific plants. At that time phenoxyacetic acids were discovered and their derivatives were carefully studied under the cover of military secrecy, both in the United States and England.

Plant hormones may be divided into five general groups. Auxins include the native indoleacetic acid, as well as the synthetic phenoxyacetic acids, notably 2,4-D (2,4-

dichlorophenoxyacetic acid). The second group is the gibberellins which are steroid-like compounds comprising over forty different structures. The third group, the cytokinins, are all N^6 substituted adenosine compounds. The fourth class is represented by a single substance, abscisic acid, which is also derived from isoprenoid units as are the gibberellins. Ethylene, the simplest of all the plant hormones, is a gas and therefore is easily spread from organs of production to organs of sensitivity.

It is interesting that of these five classes of hormones, two are generated from mevalonic acid. Ethylene is produced by the metabolism of methionine and indoleacetic acid is produced by the removal of carbon and nitrogen from tryptophane. Cytokinins, in terms of synthesis, are probably the most complex hormones. They are possibly produced from the breakdown of transfer RNA (tRNA). In the intact tRNA, the adenosine moiety of the RNA is modified by the addition of an isoprenoid group; subsequently, the tRNA is probably degraded, yielding isopentenyl adenosine, (IPA) the endogenous cytokinin.

One might reasonably imagine that an understanding of the physiological role and biochemical mechanisms of action of the plant hormones would have been elucidated quite some time ago. This of course, is in relation to what is known about the many and complex hormones in animals. Unfortunately, however, the plant hormones are not as well understood. Probably the main obstacle to the elucidation of the action of plant hormones has been the fact that they have overlapping and complementary effects on the actions of each other.

During the years between 1960 and 1970 substantial progress appeared to be made on the biochemical mechanisms of action of plant hormones. Much of this research dealt with the effect of the hormones on nucleic acid metabolism, protein synthesis and the synthesis of specific enzymes, since it was felt that the mechanism of hormone action would involve the control of the production of messenger-RNA (mRNA) and the *de novo* synthesis of enzymes. In the late 1960's the general nature of the research was radically changed because many investigators had found that the plant hormones mediated an effect on cell growth within a few minutes. This information implied a very rapid primary response mechanism of hormone action; consequently, these effects could not be mediated by changing the rate or type of synthesis of nucleic acids or of specific proteins. Many years of research have since been devoted to studies dealing with short-time growth responses to plant hormones. It now seems most likely that plant hormones regulate plant growth through both short-term and long-term growth controls.

It is fairly well understood that animal hormones first interact with the animal cell by binding to, or reacting with, a receptor site of some type within the target cell. Plant physiologists and biochemists are now beginning to consider that the action of many of the plant hormones similarly involves their binding to receptor agents which in turn amplify the action of the hormone and thereby bring about specific changes in nucleic acid synthesis, protein synthesis, enzyme activity and possibly other physiological responses, such as changes in permeability of membranes.

When a hormone is applied to a responsive plant system it brings about a specific change which results eventually in a measurable biochemical or physiological effect. Two distinct aspects of the measured effect are involved: the specific change in metabolism and the series of steps which lead to the physiological effect. Usually, the molecular interaction of the hormone at its site of action is referred to as the *mechanism* of action. The subsequent sequence of reactions leading to the physiological effect is referred to as the *mode* of action.

Therefore, by definition each hormone has its own distinctive mechanism of action even though the manifestation of the hormone mechanism may depend upon prior action by other factors. Thus, it is possible for the hormonal *mechanisms* in one plant system to lead to a series of physiological responses which may be completely different from those in a second system. This difference in *mode* of action can be brought about because the second system has more or less of another hormone, has other biochemically rate-limiting components or possesses structural and cytological differences.

AUXIN ACTIONS

Amongst plant growth regulators, the auxins have been studied the longest and have involved the greatest amount of research. This class of hormones is known to promote cell enlargement or cell elongation, a process which requires extension of the cell wall.

Commencing in 1940, the majority of studies involving auxins were begun in an attempt to determine the molecular interaction with the plant cell. Most of this research was directed toward the cell wall because it was believed that auxin action required a change in the cell wall to allow for cellular expansion. Thus, the effects of the hormone were thought either to change the cell wall deposition or to hydrolyse certain cross-linkages in cell walls making them more elastic.

This type of research was carried out for several years without achieving a clear understanding of how auxin might control such processes within the cell wall. In the 1950's a new area of research was begun. Under the direction of F. K. Skoog in Madison, Wisconsin, a report was published which showed that the auxin, indole-3-acetic acid, remarkedly affected the RNA and DNA contents of plants. Auxin increased the content of nucleic acids in tobacco tissue cultured on a sucrose-agar medium.

This increase occurred prior to the auxin-induced growth of the tissue at concentrations of IAA which were optimal for cell enlargement. For many years following this discovery much research was devoted to the general mechanism by which auxin increased the synthesis of nucleic acids.

A great share of this research was begun in J. Hanson's laboratory in Urbana, Illinois by West and Key who showed that the synthetic auxin, 2,4-D, produced a wide range of morphological and physiological changes in the hypocotyl of soybean and the mesocotyl of corn. Within 15–24 hours after 2,4-D treatment, cellular enlargement was noted concomitant with an increase in the size of the nucleus. Accompanying these changes were dramatic increases in RNA content, most of which were due to increased ribosomal-RNA. Chrispeels and Hanson suggested that auxin acts on the nucleus causing it to revert to a meristematic type of metabolism.

The role of the nucleus in such a sequence of events is of obvious importance since the nucleus has been shown to accumulate RNA in response to auxin. Beginning in the 1960's, Key began experiments on auxin regulation of nucleic acid synthesis. He and Ingle demonstrated that auxin controls the synthesis of nucleic acids other than that of ribosomal-RNA. Their data revealed that auxin causes production of nucleic acid which appears to be of the messenger-RNA type.

Subsequent experiments by O'Brien *et al.*, showed that treatment of soybean with auxin caused a large increase in chromatin-directed RNA synthesis. It was of interest to note that auxin-induced RNA synthesis produced a type of RNA which was different from the control RNA as judged by molecular size and nearest neighbour analysis. Following this area of research other studies showed that 2,4-D increased the activity of RNA polymerase.

Current research suggests that particular cytoplasmic or membrane bound factors may enhance the activity of RNA polymerase. It is hypothesized that the mechanism of action of auxin is to bind to a receptor molecule located on the plasma membrane. The receptor is then released and moves into the nucleus where it modulates the activity of the RNA polymerase. The increase in RNA polymerase activity leads to increased synthesis of messenger-RNAs, which in turn regulate, or control, the synthesis of specific proteins within the target cells.

A primary challenge is to isolate the factor which binds to, or reacts with, auxin, and determine whether this is the primary action of the hormone. Secondarily, research is needed to determine what series of events this interaction puts into metabolic play. It is likely, once the auxin has bound to the receptor molecule within the target cells, that the changes in nucleic acid synthesis cause an increased activity of the various enzymes associated with the cell wall. Many other activities which have been measured over the many years are probably secondary effects resulting from the primary action of the auxin reacting with its receptor molecule.

GIBBERELLIN ACTION

Gibberellins may be defined as compounds having a gibbane skeleton and biological activity in stimulating cell division or cell elongation or both (various structures are given in Fig. 1).

Gibberelins also may possess other biological activities such as induction of nucleic acid and enzyme synthesis. Presently there are more than forty known gibberellins which have been isolated from green plants or the fungus *Gibberella fujikuroi*. Since the late 1960's three to four new gibberellins have been identified each year. Thus, it seems that great progress is being made in the identification of naturally-occurring gibberellins; on the other hand the mechanism of action of gibberellins is still unknown. Plant physiologists are therefore confronted with a very wide array of gibberellins, several of which often occur in the same tissue, and the question arises as to their function within the cells.

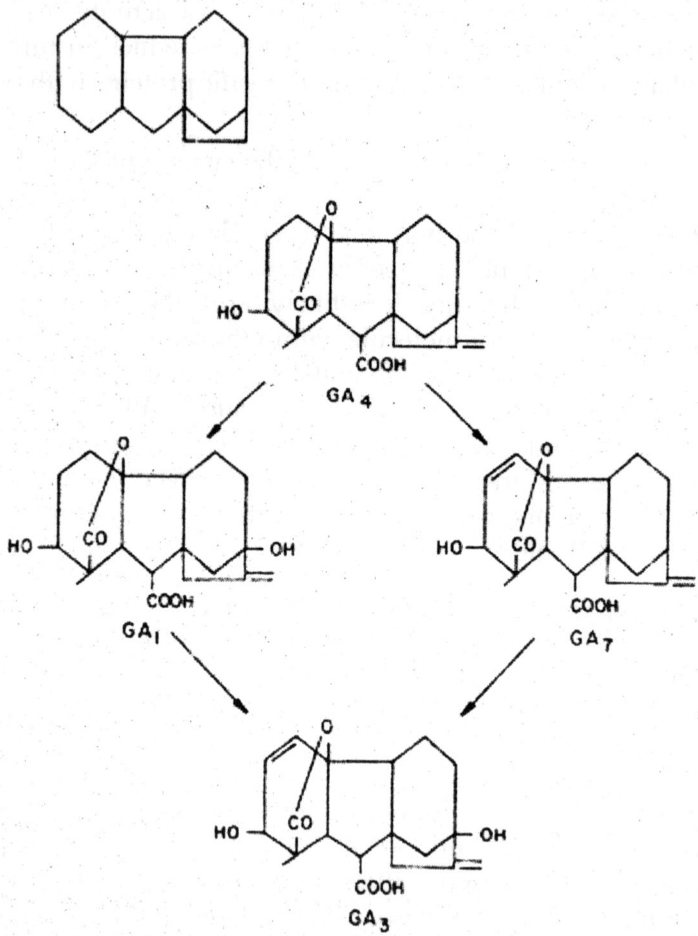

Figure 1. Structure of gibbane and the conversion of GA4 to common gibberellins.

Work beginning in the 1960's on the possible regulation of gene transcription by gibberellins was initially successful and achieved wide-spread acceptance. Later work raised many doubts, however, and at the present time, many workers in the field reject the concept that gibberellins regulate physiological processes through the control of gene transcription. Alternative mechanisms involving an action of gibberellins on lipids and membranes are now being proposed.

ACTION OF CYTOKININS

The term cytokinin has been accepted practically universally as a generic name for substances which promote cell division and exert other growth regulatory functions in the same manner as kinetin. The synthesis and testing of compounds for cytokinin activity began with the discovery of kinetin (6-furfurylaminopurine). Today there are probably at least one hundred known synthetic and native cytokinins.

Structural requirements for a high order of cytokinin activity generally include an adenine molecule with the purine ring intact and with an N^6 substituent of moderate size. One exception to the requirement for a modified purine exists, notably diphenylurea and its derivatives. Certain other substances which lack a true purine ring, such as 8-azakinetin, 6-benzylamino-8-azapurine and 6-(3-methyl-2-butenylamino)-8-azapurine are active, but each of them is less than 10% as active as its corresponding purine derivative. Substitution of either O or S for N in the N^6 position of adenine in each case results in a more than 90% loss in growth promoting activity in the tobacco bio-assay.

Cytokinins play a role in practically all phases of plant development from cell division to the formation of flowers and fruits. They affect metabolism including the activities of enzymes and the biosynthesis of growth factors. They also influence the biogenesis of organelles and the flow of assimilates and nutrients within the plant. Cytokinins defer senescence and may protect the plant against adverse environments, such as water stress. These many diverse effects presumably stem from some primary anabolic function of the cytokinins that remains to be elucidated.

Cytokinins have been found in certain transfer-RNA (t-RNA) molecules from a large number of organisms including bacteria, yeast, plants and animals. In sequence analysis the native cytokinin, isopentenyl adenine, always occurs adjacent (3' side) to the anticodon. (Fig. 2)

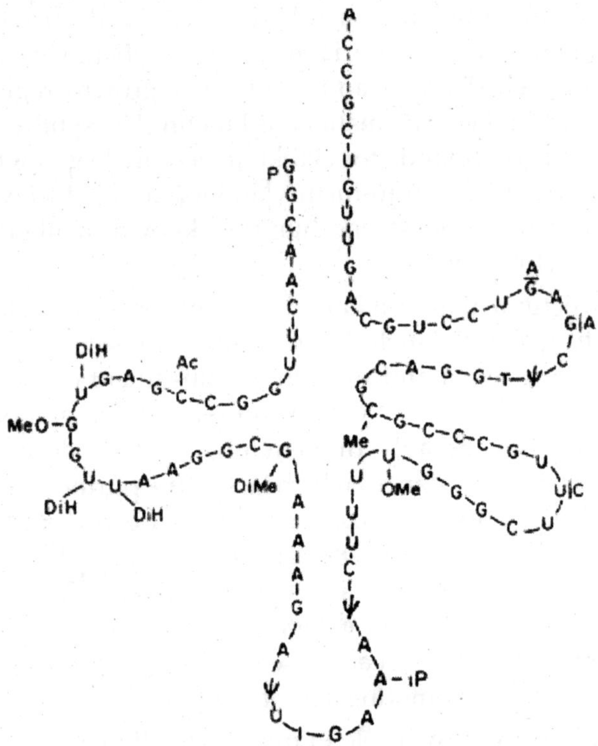

Figure 2. Diagram of the cloverleaf shape of seryl-transfer RNA.

Furthermore, the location of the isopentenyl group is required for the specific tRNA to function efficiently in protein synthesis. Results involving bacteriophage infection of bacteria indicate that cytokinin in tRNA may have an important regulatory function in protein synthesis. In soybean seedlings the application of a cytokinin (6-benzyladenine) results in increased levels of two species of leucyl-tRNAs. Although the relationship of cytokinins to tRNA may not be directly associated with the primary action of cytokinins on cellular activity, the alteration of specific tRNAs could control protein synthesis. Such a regulatory mechanism would be manifested in the control of growth and morphogenesis as the ultimate expression of the hormone.

ACTIONS OF ETHYLENE

Ethylene is the simplest hormone which regulates plant growth (Fig. 3). It is a natural constituent of plants and affects a wide array of physiological processes. For many years, however, investigations of ethylene physiology were concerned primarily with fruit ripening. With the advent of gas chromatography many experimenters have begun research on the biochemical and physiological action of ethylene.

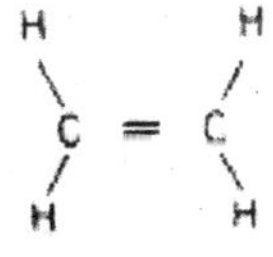

Figure 3. Structure of ethylene.

The production of ethylene by germinating seeds and seedlings suggests that the hormone may be involved in the normal regulation of growth and development. Evidence for such a proposal includes the fact that low concentrations of applied ethylene blocks photo-induced apical bud expansion and hook opening in etiolated pea seedlings. It was shown that apical tissues of etiolated seedlings are the major site of ethylene production. Light has been found to decrease the tendency of pea stem segments to produce ethylene in response to high concentration of auxin.

On the other hand, red light appears to stimulate ethylene production in dormancy. Another important role of endogenous ethylene in etiolated seedlings is the regulation of radial expansion of the pea epicotyl in a region below the apical hook. The exposure of the plant to red light or to CO_2 inhibits an ethylene-mediated increase in the diameter of the epicotyl.

Growth, flowering, abscission and fruit ripening all are affected by ethylene. It is currently popular to speculate that the mode of action of ethylene involves a mechanism which regulates some aspect of the transcription of DNA or the translation of RNA. Studies of the effects of ethylene on abscission and growth indicate sizeable changes in RNA and protein contents. There is indeed evidence of changes in the activities of catalase, peroxidase, and other hydrolases. Furthermore a study showed that exposure of soybean plants to ethylene significantly altered RNA polymerase activity associated with chromatin. As judged by nearest neighbour analysis the RNA produced from ethylene treated plants has a different base composition than that from the control. Even though ethylene at very low concentrations affects a wide array of physiological and biochemical processes in plants, much more work is required to know the mechanism of action.

ACTIONS OF ABSCISIC ACID

Abscisic acid (ABA) is a plant hormone which now ranks in importanance with the auxins, gibberellins, cytokinins, and ethylene. Interest in the physiology and chemistry of ABA has grown greatly since the structure was established in 1965. During the 1950's and early 1960's a number of laboratories were engaged in research on growth-inhibiting substances. ABA was first isolated from cotton plants and was named abscisin II by a team from Addicott's laboratory. During the same year Wareing's research team isolated an active substance from *Acer* leaves which they named dormin. Abscisin II and dormin are the same substance which is now called abscisic acid (Fig. 4).

Plant tissues of all ages appear to synthesize and inactivate ABA. The number of various plant responses affected by ABA is very large. Generally, the physiological processes are related to senescence or abscission and growth retardation or inhibition. ABA appears to act as an abscission accelerating hormone in many fruits and leaves. Furthermore, it also tends to induce dormancy in some woody plants. ABA has been shown to move from the leaves to the apical bud to bring about a dormant condition. In potato, the levels of inhibitors, including ABA, decrease during the quiescent period prior to renewed growth. ABA in extremely low concentration, moreover, prolongs the dormancy of excised potato buds.

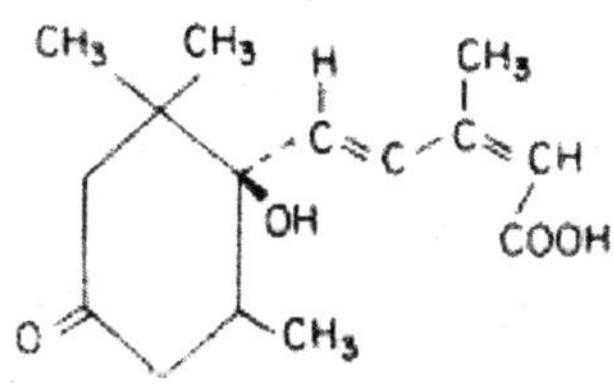

Figure 4. Structure of abscisic acid.

Recently, ABA has been shown to be involved in the responses of many plants to stress conditions. As the ABA concentration increases in the leaves the stomata close. In this way, it appears that ABA is directly involved in the opening and closing of the stomata, thus regulating the rate of transpiration. Through this mechanism, ABA appears to protect the plant through conditions of water stress or drought.

At present the mechanism of action of ABA is not clearly understood. However, the available evidence indicates that ABA affects transcription as shown by reduced activity of chromatin-associated RNA polymerases. In other cases the mechanism of action of ABA appears to involve regulation of the translation of long-lived messenger-RNA, whilst its effects on stomata probably involve the regulation of membrane permeability.

REFERENCES

Abeles F.B., *Ethylene in Plant Biology*, Academic Press, N.Y., 1973.

Jones R.L., "Gibberellins: their physiological role", *Annual Review of Plant Physiology*, 1973.

Key J.L., "Hormones and nucleic acid metabolism", *Annual Review of Plant Physiology*, 1969.

Runeckles V.C., Sondheimer E. & Walton D.C. (eds), "The Chemistry and Biochemistry of Plant Hormones", *Recent Advances in Phytochemistry*, Academic Press, 1974.

Varner J.E. & Johri M.M., "Hormonal control of enzyme synthesis", In *Biochemistry and Physiology of Plant Growth Substances* (eds. F. Wrightman & G. Setterfield), The Runge Press Ltd. Ottawa, 1968.

Bibliography

Abeles F.B., *Ethylene in Plant Biology*, Academic Press, N.Y., 1973.

Albersheim P., "Substructure and function of the cell wall", In *Plant Biochemistry* (Ed J.E. Varner), Academic Press, New York, 1965.

Allen, R. D., N. Strömgren-Allen, and J. F. Travis, "Video-enhanced contrast, differential interference contrast (AVEC-DIC) microscopy", *Cell Motility*, 1981.

Allende J.E., "Protein biosynthesis in plant systems", *Techniques in Protein Biosynthesis*, vol. 2 (eds. P.N. Campbell & J.R. Sargeant), pp. 55-100. Academic Press, London and New York, 1970.

Anderson W.F., Bosch L., Gros F., Grunberg-Manago M., Ochoa S., Rich A. & Staehelin Th., "Initiation of protein synthesis in prokaryotic and eukaryotic systems", *FEBS Letters*, 1974.

Anderson W.P. (ed.), *Ion Transport in Plants*, Academic Press, London and N.Y., 1973.

Arntzen C.J. & Briantais J.M., "Chloroplast structure and function", In *Bioenergetics of Photosynthesis*, (ed. Govindjee). Academic Press, New York, 1975.

Aspinall G.O., "Carbohydrate polymers of plant cell walls", In *Biogenesis of Plant Cell Wall Polysaccharides* (Ed. F. Loewus) pp. 95-115, Academic Press, New York, 1973.

Barthlott, W.: Epidermal and seed surface characters of plants: systematic applicability and some evolutionary aspects. *Nordic Journal of Botany*, 1981.

Bateman D.F. & Basham H.G., "Degradation of plant cell walls and membranes by microbial enzymes", In *Physiological Plant Pathology* Vol. I (Ed. P.H. Williams & R. Heitifus) Springer-Verlag, Berlin, 1975.

Beck, C. B., R. Schmid, G. W. Rothwell, "Stelar morphology and the primary vascular system of seed plants", *Botany Revolution*, 1982.

Beevers H., "Glyoxysomes of castor bean endosperm and their relation to gluconeogenesis", *Annals of the New York Academy of Sciences*, 1969.

Behnke, H.-D., W. Barthlott, "New evidence from the ultrastructural and micromorphological fields in angiosperm classification", *Nordic Journal of Botany*, 1983.

Bierhorst, D. W., *Morphology of vascular plants*, New York: The Macmillan Comp, 1971

Bonner W.D., Jr., "Mitochondria and plant respiration", In *Phytochemistry*, vol. 3 (ed. L.P. Miller). Van Nostrand Reinhold, 1973.

Boulter D., "Protein synthesis in plants", *Annual Review of Plant Physiology*, 1970.

Bowen, W. R., *Experimental Cell Biology*, New York: The Macmillan Comp., 1969

Branton D. & Deamer C.W., *Membrane Structure*, Springer-Verlag, New York/Wien, 1972.

Bryant J.A., "Nuclear DNA", In *Molecular Aspects of Gene Expression in Plants* (ed. J.A. Bryant), pp. 1-51. Academic Press, London, 1976.

Calvin M. & Bassham J.A., *The Photosynthesis of Carbon Compounds*, Benjamin, New York, 1962.

Carlquist, S., *Comparative plant anatomy*, New York: Holt, Rinehart and Winston, 1961

Cheadle, V. I., and N. W. UHL, "Types of vasoular bundles in the Monocotyledonae and their relation to the late metaxylem conducting elements", American Journal of Botany, 1984.

Cheadle, V. I., "Research on xylem and phloem—Progress in fifty years", *American Journal of Botany*, 1956.

Cormack, R. G. H., "The development of the root hairs by Elodea canadensis", *New Phytologist*, 1937.

Eisenberg M. & McLaughlin S., "Lipid bilayers as models of biological membranes", *Bioscience*, 1976.

Esau, K., *Anatomy of seed plants*, New York: J. Wiley and Sons, Inc., 1960

Fahn, A., *Plant anatomy*, Oxford: Pergamon Press, 1967

Foster, A. S., "Comparative morphology of vascular plants", San Francisco: W. H. Freemen and Comp., 1959

Frey-Wyssling, A., K. Mühlethaler, *Ultrastructural plant cytology*, Amsterdam: Elsevier Publ. Comp, 1965

Gibbs M., "Photorespiration, Warburg effect and glycolate", *Annals of the New York Academy of Sciences*, 1969.

Goddard D.R. & Bonner W.D., Jr., "Cellular respiration", In *Plant Physiology*, vol. 1A (ed. F.C. Steward). Academic Press, 1960.

Grierson D., "RNA structure and metabolism", In *Molecular Aspects of Gene Expression in Plants* (ed. J.A. Bryant). pp. 53-108. Academic Press, London, 1976.

Hall, J. L. (ed.), *Electron microscopy and cytochemistry of plant cells,* Amsterdam: Elsevier/North Holland, 1978.

Haselkorn R. & Rothman-Denes L.B., "Protein synthesis", *Annual Review of Biochemistry,* 1973.

Hearle, J. W. S., J. T. Sparrow, and P. M. Cross (eds.), *The use of scanning electron microscope,* Oxford: Pergamon Press, 1972

Heber U., "Metabolite exchange between chloroplasts and cytoplasm", *Annual Review of Plant Biology,* 1974.

Heywood, V. H. (ed.), *Scanning electron microscopy,* New York, London: Academic Press 1971.

Jackson W.A. & Volk R.J., "Photorespiration", *Annals of the New York Academy of Sciences,* 1970.

Jones R.L., "Gibberellins: their physiological role", *Annual Review of Plant Physiology,* 1973.

Keegstra K., Talmadge K.W., Bauer W.D. & Albersheim P., "The structure of plant cell walls III. A model of the walls of suspension-cultured sycamore cells based on the interconnections of the macromolecular components", *Plant Physiology,* 1973.

Key J.L., "Hormones and nucleic acid metabolism", *Annual Review of Plant Physiology,* 1969.

Kiermayer, O (ed.), "Cytomorphogenesis in plants", (Cell Biology Monographs 8) Wien - New York: Springer Verlag, 1981.

Kirk J.T.O. & Tilney-Basset R.A.E., *The Plastids,* Freeman, London and San Francisco, 1967.

Kornberg A., *DNA Synthesis,* W.H. Freeman & Co., San Francisco, 1974.

Lamport D.T.A., *Cell wall metabolism,* Annual Review of Plant Physiology, 1970.

Lehninger A.L., *The Mitochondrion: Molecular Basis of Structure and Function.* W.A. Benjamin, Inc., 1964.

Lewin B., "Gene Expression", Volume 2, *Eukaryotic Chromosomes,* J. Wiley & Sons Ltd., London, 1974.

Lichtscheidl, I. K., D. G. Weiss, "Visualization of submicroscopic structures in the cytoplasm of Allium cepa inner epidermal cells by video-enhanced contrast light microscopy", *European Journal of Cell Biology,* 1988.

Lucy J.A., *Lipids and membranes,* FEBS Lett. 40, S105-S111, 1974.

Mans R.J., "Protein synthesis in higher plants", *Annual Review of Plant Physiology,* 1967.

Munn E.A., *The Structure of Mitochondria*, Academic Press, 1974.

Parish J.H., *Principles and Practice of Experiments with Nucleic Acids*, Longman, London, 1972.

Park R.B. & Sane P.V., "Distribution of function and structure in chloropalst lamellae", *Annual Review of Plant physiology*, 1971.

Rae P.M.M., "The distribution of repetitive DNA sequences in chromosomes", In *Advances in Cell Biology and Molecular Biology* (ed. E.J. Dupraw), Vol. 2, pp. 109-49. Academic Press, New York, 1972.

Raven J.A. & Smith F.A., "Significance of hydrogen in transport in plant cells", *Canadian Journal of Botany*, 1974.

Runeckles V.C., Sondheimer E. & Walton D.C. (eds), "The Chemistry and Biochemistry of Plant Hormones", *Recent Advances in Phytochemistry*, Academic Press, 1974.

Sato S. (ed.), "Mitochondria", In *Selected Papers in Biochemistry*, vol. 10. University Park Press, 1972.

Skipworth, J. P., "The primary vascular system and phyllotaxis in Hectorella caespitosa", *New Zealand Journal*, 1962.

Slater E.C., Zaniuga Z. & Wojtczak L. (eds.), *Biochemistry of Mitochondria*, Academic Press, 1967.

Stewart P.R. & Letham D.S. (eds.), *The Ribonucleic Acids*, Springer-Verlag, Berlin, 1973.

Tanaka, K., "Scanning electron microscopy of intracellular structures", *Intern. Rev. Cytology*, 1980.

Tolbert N.E., "Microbodies-peroxisomes and glyoxysomes", *Annals of the New York Academy of Sciences*, 1971.

Trebst A., "Energy conservation in photosynthetic electron transport of chloroplasts", *Annual Review of Plant physiology*, 1974.

Turrell, F. M., "Leaf surface of a twentyone-year-old catalpa tree", *Iowa Acad. Sci. Proc.*, 1934.

Varner J.E. & Johri M.M., "Hormonal control of enzyme synthesis", In *Biochemistry and Physiology of Plant Growth Substances* (eds. F. Wrightman & G. Setterfield), The Runge Press Ltd. Ottawa, 1968.

Yamaguchi, Y., R. Nagai, "Motile apparatus in Vallisneria leaf cells", I. Organization of microfilaments, *Journal of Cell Science*, 1981.

Zalik S. & Jones B.L., "Protein biosynthesis", *Annual Review of Plant Physiology*, 1973.

Zelitch I., *Photosynthesis, Photorespiration and Plant Productivity*, Academic Press, New York, 1971.

Zimmermann, M. H., P. B. Tomlinson, "The vascular system of monocotyledoneous stems", Botany Gazatte, 1972.

Index